Research Ethics for Scientists

A Companion for Students

Second Edition

C. Neal Stewart, Jr.
University of Tennessee
Knoxville, TN, USA

Registered Offices
John Wiley & Sons, Inc., 111 River Street, Hoboken, NJ 07030, USA
John Wiley & Sons Ltd, The Atrium, Southern Gate, Chichester, West Sussex, PO19 8SQ, UK

For details of our global editorial offices, customer services, and more information about Wiley products visit us at www.wiley.com.

Wiley also publishes its books in a variety of electronic formats and by print-on-demand. Some content that appears in standard print versions of this book may not be available in other formats.

Library of Congress Cataloging-in-Publication Data
Names: Stewart, C. Neal, Jr., author.
Title: Research ethics for scientists : a companion for students / C. Neal
   Stewart, Jr., University of Tennessee.
Description: 2nd edition. | Hoboken, NJ : Wiley, 2023. | Includes
   bibliographical references and index.
Identifiers: LCCN 2023002864 (print) | LCCN 2023002865 (ebook) | ISBN
   9781119837886 (paperback) | ISBN 9781119837893 (adobe pdf) | ISBN
   9781119837909 (epub)
Subjects: LCSH: Research—Moral and ethical aspects. |
   Scientists—Professional ethics.
Classification: LCC Q180.55.M67 S76 2023 (print) | LCC Q180.55.M67
   (ebook) | DDC 174/.95—dc23/eng20230410
LC record available at https://lccn.loc.gov/2023002864
LC ebook record available at https://lccn.loc.gov/2023002865

Cover Design: Wiley
Cover Image: © Sean Xu/Shutterstock, Rosamar/Shutterstock

Set in 10/13pt BemboStd by Straive, Pondicherry, India
Printed and bound by CPI Group (UK) Ltd, Croydon, CR0 4YY

C9781119837886_030723

To Sara, the love of my life and the most ethical person I know.

Contents

Contents

Contents

Preface

The second edition of this book was written in the midst of the chaos that was the COVID-19 pandemic accompanied by my own deliverance to new abodes on the banks of the Tennessee River: changes and more changes ruled these years.

From my new digs I watched the river flow along with flotsam and jetsam, which included all manner of biomass, architectural remnants, and the broken products of human ingenuity. Also all sorts of boats comprised my daily entertainment: from pontoons, cruisers, runabouts, the occasional sailboat, kayaks, huge barges, to the dozens of bass boats launching toward the day's catch in search of winning an elite fishing tournament.

In my head played Bob Dylan's 1971 breakthrough hit that escorted him into the realm of rhythm and blues a la Leon Russell: "Watching the River Flow." He sings:

> People disagreeing everywhere you look
> Makes you wanna stop and read a book
> Why only yesterday I saw somebody on the street
> That was really shook
> But this ol' river keeps on rollin', though
> No matter what gets in the way and which way the wind
> does blow
> And as long as it does I'll just sit here
> And watch the river flow

I am certain that many people view scientific research in the same vein as winning a fishing tournament: a competition that values speed, cunning, strategy, and persistence. I suppose you can count me as one of them. The majority of professional anglers play by the rules. Still, cheaters exist. One particular stunt comes to mind. A couple of anglers had a difficult day on the water and decided to stuff their few landed fish full of lead to escalate the weight of their catch. They apparently got hooked on this scam since they were not immediately detected. And so, they became repeat offenders. Unfortunately for them, they were eventually trapped in a net of their own deceit. And, the honest anglers were incensed at the lack of integrity invading their sport. Telling a fish story in one thing, but blatant cheating was intolerable.

The same is true in science: cheating can never be acceptable in research. Sad to say, one of the main motivations for updating Research Ethics for Scientists with this second edition is that dishonest researchers have found numerous new ways to falsify, fabricate, and plagiarize. I tried to cover all the dastardly innovations in the book as well as the defenses. Despite the negatives, I am convinced that the overwhelming majority of scientists are honest to the bone. They continue to do the hard work of scientific research to benefit humankind.

Thus, the river of science keeps on rollin' even though it is sometimes riddled with flotsam and jetsam and the occasional bad player.

I hope this does make you wanna stop and read the book.

Louisville, Tennessee, USA
March 2023

Acknowledgments

I am greatly indebted to many people who played an important role in this second edition. My able assistant and lab manager Cassie Halvorsen keeps my professional world organized. She also helped with finding and managing figures and tables. Thanks to Lannett Edwards who has co-taught the graduate course of Research Ethics each fall for the past 17 years. I continue to learn from her as well as the hundreds of students who have taken the course. Thanks to my lab group who challenge me in all the right ways and keep me honest each day. Of course, there is the occasional ones who challenge me in not-so-helpful ways, but I manage to learn from them too. I really appreciate Daniel Anderson and Bob Langer for providing updated interviews for the book. It still amazes me that with their busy schedules, they are so quick to respond to my questions. Thanks to people who gave me permission to use their material in the second edition. I am grateful to all of them, especially Ivan Oransky of Retraction Watch and Michael Lauer of Open Mike fame at the NIH. I continue to appreciate my mentor at the University of Tennessee: Oz Augé. Even in retirement he continues to send me nuggets of ethical gold via email. Finally, thanks to Rebecca Ralf and the great team at Wiley for making the second edition a reality.

To God be the glory.

About the Companion Website

This book is accompanied by a companion website.

www.wiley.com/go/stewart/researchethics2

This website includes:
✓ Case Studies
✓ Figures from the book as PPTs

Chapter 1

Research Ethics: The Best Ethical Practices Produce the Best Science

---

ABOUT THIS CHAPTER

- Research science is increasingly complex with pitfalls and temptations.
- Global competition and cooperation will likely change the face of science in the future.
- Science is an iterative loop of ideas, funding, data, publication, leading back to more ideas and research.
- Ethics can be a guide toward best practices.
- Best scientific practices lead to the best science results and discoveries.
- Best practices and mentorship give rise to the best scientists.

It is increasingly difficult to be a research scientist. The number and complexity of rules, electronic forms, journals and publishing, and university regulations are ever-growing. The competition for funding is often ruthless, and the criteria exacted to warrant publication in good journals also seem to be on the rise. Indeed, not just the pressure to publish, but the pressure to publish the right papers in the right journals is also increasing. Nominally, the preparation of proposals and publications has been ostensibly made simpler by computer technology, yet the potential for real- and faux-research productivity has also been enabled by computers. Technology is a double-edged sword, enabling high levels of knowledge creation as well as enabling research fraud and shoddy science. Thus, ethical dilemmas seem to

---

Research Ethics for Scientists: A Companion for Students, Second Edition. C. Neal Stewart, Jr.
© 2023 John Wiley & Sons Ltd. Published 2023 by John Wiley & Sons Ltd.
Companion website: www.wiley.com/go/stewart/researchethics2

be appearing at an increasingly rapid pace, with research misconduct regularly being the subject of news articles in *Science, Nature,* and *The Scientist.* Even people who do not keep up with science news are familiar with breakthroughs in science and controversial developments such as CRISPR babies. While the most notorious cases of misconduct have occurred in higher-profile fields of science, such as physics and biomedicine, it is clear that no area of science is immune to unethical behavior (Judson 2004; Ritchie 2020).

We live in a "multi" world. Multitasking, multidisciplinary work, and multiauthored papers, to name a few, are ingrained in the fabric of science culture and certainly multi-multi is expected in order to succeed and move up the scientific ranks. The isolated small laboratory with the lone professor and few staff (see Weaver (1948) for a perspective) has given way to larger labs interacting in complex collaborations in interdisciplinary science. Complex relationships are accompanied with tough decisions regarding authorship, dicing the funding pie, and how to treat privileged data, and immense amount of data at that, which are shared (or not) and curated in useful and meaningful ways (or not). In all this mix, the temptation to cheat, cut corners, and misbehave seems to be at its zenith for scientists wishing to compete at the highest levels of science, become tenured, and then become rich and famous. Well, ok, realistically, most of us are challenged to name more than a handful of scientists who ever became rich or famous. Of course, one alternative to honest competition and competence, as seems to be the case for some scientists, is to con their way to the top. Cheating is front page news in business, politics, and sports news alike. Perhaps a bigger problem to outright fraud is cutting ethical corners. Thus, we have an apparent paradox – the antithesis of this chapter title – that the best (or highly rewarded) science is compromised with seemingly endless ethical issues. Whereas the lone professor and his or her graduate student worked in simpler and more linear paths in the past, modern science seems far too convoluted for its own good (Munck 1997). How can we win? How can sound science prevail in the face of all the obstacles?

If the situation is not complicated enough, it seems that there are growing concerns about the abuse of graduate students and postdocs by their mentors. Some senior scientists feel that coercion, micromanagement, and general overbearance of their trainees are effective means to ensure high productivity. While research misconduct garners headlines, causing all sorts of angst upon university administrators, it might be the case that defective mentorship is actually a much weightier problem than outright cheating (Shamoo and Resnik 2003). But is it possible that these two problems could be interconnected? Mentorship is a perennial hot topic in science that has spawned cottage industries, self-help books, and strategizing among faculty members and university administrators alike. Everyone knows that finding good mentors is crucial for the young (and sometimes not-so-young) scientist wishing to be propelled into a sustainable career in the academic world of research and teaching or the private sector of research. Mentors, after all, know the unwritten rules of science and can share these with their trainees. Mentors are responsible to explain how these rules are intermeshed with research ethics and advice on best practices. Mentors should help their students and postdoctoral trainees fulfill their dreams (should their dreams involve being a scientist). Indeed, these features define good mentorship. Bad mentors can shatter dreams and stagnate their trainees' careers. But perhaps even the best mentoring is not effective to deter certain research misconduct.

Research misconduct is a major threat to science. As much as some scientists wish to point fingers at politicians and the public as the principal bad players responsible for lack of appreciation and funding that science deserves, I think the real enemy is within our own ranks. Indeed, Brian Martin (1992) maintains that modern science, the "power structure of science," is to blame for much misrepresentation in research. Essentially, publishable data (indeed, stories) must be novel to be publishable in the sorts of journals that scientists need to publish in. According to Martin scientists are not allowed to "tell it like it is" and must "sell" publishable stories (he calls them "myths"). Nonetheless, blatant untruths in publishing are typically rooted out as research misconduct. Papers found to contain false information – created either by misconduct or honest error are typically retracted. Research misconduct is insidiously damaging to the credibility of science and scientists in society since it erodes trust – not only in the individual researchers but in the system of science itself. Self-patrolling the profession from within is critical to reverse this damaging trend; the major pinch points for detecting research misconduct are when grant applications are submitted and when manuscripts are assessed at the editorial level and peer reviewed.

3

The ethical dilemmas in data collection, collaboration, publication, and granting are likely to become even more complex and vexing in the future. More than ever, graduate students and postdocs must master more techniques, technologies, and concepts in order to become and stay competitive in science. At the same time, scientists must generate good ideas and raise increasingly scarce funds to make their research a reality. Global competition from scientists in rapidly developing countries, especially in Asia, is a new fact of life for the researchers in the West, who were quite accustomed to the deck stacked in their favor. Researchers in China, India, countries in the Middle East, and in other rapidly developing countries are enjoying increased levels of new funding and increased status in the world of science. These new resources are coupled with even higher government and institutional expectations – not only for results and publications – but groundbreaking publications in the most prestigious journals. From East to West, being a practicing scientist is certainly not getting any easier. The picture is not all doom and gloom, however. Honestly, I can think of no more exciting time to be a scientific researcher than today with the booming innovations and opportunities to be found around every corner. For example, between the publication of the first edition of this book (2011) and the second edition, I essentially reset my lab to perform synthetic biology research with new funding sources and collaborators. This transformation has led to new facilities and innovations. Such innovations are also enabled by our ability to connect with other scientists and stakeholders across the globe nearly instantaneously these days. Certainly, the positive science news outweighs the negative news and complications, but there is great consensus among scientists and others that the science system, while considered to be self-correcting, can go awry.

It was around 2006 and 2007 that I became convinced, for all the above reasons (as well as others discussed later in this chapter), that a new course at my university needed to be taught on research ethics to graduate students. Thus, I embarked on learning a lot more about ethics, research integrity, and the many topics that touch responsible conduct in research. After a couple years teaching this new graduate course, I decided that a book of this sort could be helpful to support it, but also as a general help to young scientists just starting their research careers. A few years later, the first edition of this book was published. Over a decade later, I found that a lot had changed (well, mainly, people had come up with new ways to cheat), and a second edition was due.

This book could be viewed as part guidebook, part virtual mentor, and part friendly polemic that should be helpful in addressing pragmatic problems that all research scientists experience. While virtual mentoring was part of my motivation, to substitute any book for finding a real mentor would be a mistake, which is one main reason a couple chapters on mentorship are included. This book is on research ethics, users' guide to success in science by following the rules that scientists largely agree are requisites for success. This book will not focus on greater issues of morality or bioethics; those are vastly different topics than the one we are embarking on here.

And with that, I'll state up front that I do not have all the answers. I think I do ask most of the pertinent questions, but like most things in life, asking the questions is a good bit easier than answering them. One of my main goals in asking the questions is to enable the readers to judge themselves with regards to best practices. When I started in science, I expected that there would be clearly illuminated a singular correct way to do experiments, analyze the data, and write up the papers. It did not take long to learn that this was not the case, and indeed, I judged myself then and ever-frequently now in how I could improve. Science is very creative and individualistic. There are many ways to answer scientific questions, and many ways also to go wrong. That is not to say that we cannot learn from our mistakes and at least not doom ourselves in repeating the same mistakes over and over again.

So, I urge the reader to think about the questions and the answers. I further ask readers to think about opinions expressed here, especially analyzing the case studies for current and future action where applicable, so that an individualistic way forward is clearly seen for each scientist embarking on the individualistic and exciting journey that is research science. I've found it valuable to discuss topics presented in the book with colleagues and mentors. If the topics in this book are discussed more widely in labs, hallways, and classrooms, then the best ethical practices will be advanced throughout fields of science. After I began teaching about responsible research conduct and practices, I found the new lively hallway discussions about various topics related to our course content was proof positive that our new effort toward promoting best practices was worthwhile.

Judge yourself

✓ Why are you interested in research ethics?
✓ What are your motivations for pursuing research?
✓ What ways are these motivations synergistic or antagonistic with one another (research ethics vs. research)?

Morality vs. ethics

What is the difference between morality and ethics? If morality is the foundation that ethics is built upon, research ethics is the top floor that is visible from the air. The moral foundation often has religious or spiritual ingredients and is engrained in substance that is far beyond the scope of this book. Ethics is sort of practical morality or professional morality that enables fair play in research. If we think of problems not so much as in terms of right and wrong, but in terms of ought and ought not, then I think we understand how to parse morality vs. ethics. Many people are uncomfortable discussing morality, religion, and politics. In contrast, most scientists are happy to share their opinions on ethics of their fields and science in general. Most experienced scientists understand standard practices in their fields and can recognize deviations from standard practices. If we move to a higher level that encompasses all fields of science – the big picture – then there is a general agreement of standard practices among scientists. The big picture of items included in this book includes research misconduct (it is bad), mentorship (it is critical), publication ethics and authorship (it can be tricky at times but science must be published), data integrity, and preservation (without it we are sunk). One way to think about research ethics is in terms of best practices in conducting all aspects of research science – to maximize benefits and minimize harm. A very important ethics concept is nonmalfeasance: doing no harm (Barnbaum and Byron 2001). While the definitions and delineations on research ethics might seem a bit squishy, let us keep in mind that there is plenty of room for opinion. Indeed, this book includes topics that are often debated, such as peer review, grants and externally funded research, and conflicts of interest.

This book is about practical research ethics as opposed to the theoretical ethics that may be of interest to a philosopher. This book is for scientists. This book is about integrity in performing research. Summed up, this book is about scientific integrity and the scientist's role in preserving it.

For our purposes here, I view this book as mainly about how to be a successful scientist. It can easily be argued that philosophers have thought about ethics much longer (e.g., Plato and other ancient Greek philosophers) than scientists have thought about science (a word not coined until the 1800s (Shamoo and Resnik 2003)). There are many viewpoints that philosophers have taken to conceptualize ethics. A few of these are utilitarianism, deontology, and virtue ethics.

Utilitarianism is an example of teleological theory, which is based on outcomes rather than process. Utilitarianism seeks to do the most good for the most people; it is important to consider others and not just yourself. The utilitarian essentially does cost–benefit analysis to guide a person's path and decisions, and one that is widely implemented these days as a thought process (Barnbaum and Byron 2001).

Deontology is the ethics of duty. It strives to universalize rules that apply to everyone in guiding actions. One example here is the Golden Rule (or the rule of reciprocity), which is stated as, "Do unto others as you'd have them do unto you." "Morality as a public system" (Gert 1997, p. 24) applies to research ethics in that all scientists know the rules to be followed and is not irrational for the people who agree to participate in the system.

Virtue ethics focuses on living the good life. In this system, a person is guided to do what a virtuous person should do. Similar with the other two systems above, virtue ethics considers the potential for harm and avoids doing things to harm others, as that is what the virtuous person ought to do.

A last self-centered way to look at ethics is through the eyes of egoism (Comstock 2002). Egoism states that a person ought to do what is in their own self-interests. If a scientist wants to have a long and fulfilling career, then they should follow the rules and perform the best science. It is also in my own self-interest, especially in the long run, to care about others and tell the truth in science.

As a scientist, it is difficult for me to actually decide which of these various systems is most effective. To me, they all point to the same guide for behavior and context. If we mash them up, a virtuous scientist will seek the truth for the better good of humanity in following the rules that most scientists agree upon because it serves the self-interest of individual scientists. Scientists, by definition, should desire to maximize benefit and minimize harm (normative principles).

Onward and upward

This book will be about the best practices in all the major areas of research management and practice that are common to scientific researchers, especially those in academia. Aimed toward helping the scientist in formative career stage, it will critically examine the key areas that continue to plague scientists, both young and old.

The book is arranged into functional themes and units that every experienced scientist recognizes as crucial for sustained success in science: ideas, people, data, publications, and funding. For example, relative to "ideas" there will be chapters on plagiarism, credit, and fairness. Note that there is some overlap between topics (e.g., plagiarism relates also to publications), but the book seeks to integrate topics into a structure that should help students and professional scientists see the interconnectedness of components leading to successful research. Herein, I will acknowledge my own opinions and biases and weaknesses and frailties. In our research ethics course, my co-teacher and I argue (discuss) facts and opinions. If we accept that there is plenty of room for difference of opinion, then ethics discussions are a lot of fun and help clarify the way we think about doing science. Of course, it also helps to differentiate between facts and opinions.

Inauspicious beginnings

Other than the standard morality lessons about right and wrong that most children are taught, I do not remember enjoying any formal teaching on professional ethics. Well, there were the standard mandates against plagiarism and cheating on tests. I had a bit of immersion in responsible conduct of research when I was selected to be the chief justice of my university's graduate honor system (and I really do not know why I was nominated or how I was selected). For the three years during my PhD program, I presided over panels that heard cases about graduate student plagiarism, data fabrication, and falsification. I suppose I learnt the rules of science by observing the real-life cases of students breaking the rules. Even then, I reasoned that everyone valued common sense ethics and there was no need to understand the details other than do not break the rules. I viewed my "chief justice gig" – which waived my university tuition and fees and gave me a faculty parking pass – as having little to do with my own PhD research. I had compartmentalized ethics from my scientific interests. In my mind, this singular focus of research during my graduate program was by necessity. I had found myself so far over my head and out of my comfort zone in science with the main need to learn so much so fast, that it took every drop of energy I could muster, especially in the early part of graduate training, to keep from drowning. Even then, at times, I felt I was floundering in my classes and research. I think I would have considered any training or discussion about ethics, best practices in science, or even how to *be* a scientist a real distraction from science itself. How wrong I was!

Let us imagine a mechanical engineer who is fascinated with cars. The engine design, drivetrain, tires, chassis, brakes, the whole thing is an obsession. Now after studying the theory of everything automotive, our ambitious engineer designs and builds a fully functional 500 hp machine that is capable of going 0–60 mph in less than four seconds. And after all these years, our engineer will now finally drive his first car – ever – his first car being the one of his own design. Unfortunately, before taking the wheel he never learnt the rules of the road. He does not know what that octagonal sign means, whether to drive on the right or left side, and let us not even think about common courtesy. No, our engineer considered all these things to be a distraction from what was really important – the car itself – the engineering. A disastrous crash and the destruction of the beautiful work of motoring machinery is highly likely. Sad to say, the unpleasant result could have been avoided by a short course on how to drive while sharing the road with others.

While this might seem like a ridiculous example, it illustrates how many young scientists, myself included, approach learning science and being a scientist, seemingly by osmosis. One might argue that our automotive engineer would gradually learn the traffic laws and the accepted motoring behavior, maybe even from a good, personalized driving instructor over time. But how much damage could be done in the meanwhile? As more and more students come into my lab and leave as budding scientists, I've become thoroughly convinced that learning best ethical practices earlier rather than later in a research career results in a big payout to both the scientist and the science itself. There is merit to having a driving course and a handbook.

How science works

The illustration below summarizes the flow of science, at least how it is currently practiced, with all of its necessary components. Science is actually a reiterative loop in which successes beget successes and failures cause the research loop to be broken. One of the primary drivers for success (completed and reiterative loop) or failure (broken loop) is scientists themselves. Having the best trained people who are eager to do best practices are at the heart of all successful science (Figure 1.1).

For the sake of discussion, we will designate a spot in the loop as the logical endpoint: publications. The end product of science is actually new knowledge, which must be canonized as peer-reviewed journal articles. Although there are other legitimate outlets for knowledge dissemination, such as presentations in professional meetings, books, book chapters, patents, and oral histories, the "gold standard" for credible science is peer-reviewed journal articles. This has largely been the case since 1660, when the first journal, the *Philosophical Transactions of the Royal Society*, was published.

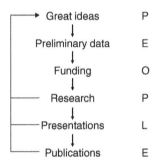

Figure 1.1 The flow of research. Research ultimately starts with great ideas and ends with publication of the research; i.e., new knowledge. Ethics is intertwined with the various steps in research and integrated with the people involved.

In most cases, a science paper is built on data from well-designed experiments that test hypotheses. While professors might likely have a hand in designing experiments and formulating hypotheses, it is graduate students, postdocs, and other bench scientists who actually collect and analyze data and do most of the writing. Actually doing science from inception to publication is the rare luxury that few tenured professors enjoy today. While the old-professor-in-the-white-lab-coat myth continues to live in popular culture, professors are producing fewer and fewer data with their own hands in the lab; in the grand universe of data, the professor-collected data is miniscule.

That is because they are busy writing grant proposals and performing administrative duties! Writing the second edition of this book was first delayed by COVID-19 and all the administrative hoops I jumped through to keep my lab running through the pandemic. When I was ready to write, I found that I had a once-in-five-years opportunity to simultaneously participate in six grant proposals, one of which I led. Last week I attended a research conference and submitted two research studies for publication. Grant proposals (whether as a principal investigator (PI) or co-PI) are necessary to fund research. The PI is the scientist taking the lead in the proposal, and a co-PI is a key person helping to write a multi-person team proposal. In most colleges and universities, the only scientists who are typically paid from "hard" funding, that is from university-level funding, are professors (and then again, in US medical schools, even professors are responsible to raise much of their own salaries). Ironic is the world of science in that the least productive people, data-wise, are the ones who have a tenure system to protect their employment status and salary stability. Everybody else – the ones doing the work – are typically on "soft" (grant) money. Why the disparity? A partial explanation is that faculty teach and are paid from university tuition income, but it is widely known that professors who attract a lot of grant funding and those with high research productivity (read, publications) are the scientists who are most esteemed in science and by higher education administration. In science, these professors are typically the scientists with the highest statures and salaries. Again, why? They are the ones who enable the funding of science to collect the data to publish the papers. Famous papers containing groundbreaking science in turn yield status to institutions (and more money), thus the financial circle is completed. Universities successful in research have greater reputation and funds enabling them to get even richer, hire more faculty members, and continue the trend of the rich getting richer. See the box describing how Vannevar Bush enabled the establishment of the American system of funding basic science that has been copied throughout the world. This short history lesson goes a long way to help understand why the science research system exists and also, to a degree, why it works.

From the endless frontier to NSF and DARPA: how science is funded by the US government

Most of the scientific research at universities and research institutions is funded by government agencies. In the United States, nonmedical basic research is funded largely by the National Science Foundation (NSF), which was founded in 1950. The NSF's beginning can be traced back to a 1945 report to President Harry Truman authored by Vannevar Bush entitled "Science – the Endless Frontier" (Bush 1945). The report was the product of scientific breakthroughs experienced by the United States during World War II, which prominently included the Manhattan Project that developed the first atomic weapons. The point of the report was that during peacetime scientific research was actually an endless frontier of discovery. Since brilliant scientists have brilliant ideas, Bush reasoned that the formulation about topics for research ought to be made by scientists themselves and publicly funded. Bush made the point that both fighting disease and creating jobs would come from basic scientific research. The report made the recommendation that the government should provide "full support of some over-all agency" to fund university researchers. Thus, after some political wrangling, the National Science Foundation (NSF) was created. It grew to contain various directorates tied to scientific disciplines that requested research proposals from faculty in universities. Thus, the NSF became an important research model for funding basic, investigator-driven research.

In 1958, quite a different sort of funding agency was founded in the United States: the Advanced Research Project Agency, which became the Defense Advanced Research Project Agency (DARPA) in 1972. Unlike the NSF, which funds relatively "safe" incremental science projects that will likely lead to fundamental new knowledge, DARPA funds "high risk-high reward" projects based on technological breakthroughs. One such breakthrough was the creation of the ARPANET in 1969, which was the principal precursor to the internet. Like the NSF, DARPA has been the model in the United States and abroad for funding relatively risky research, which, if successful, has a big payoff. Some examples in the United States are ARPA-E (energy), ARPA-H (health), and IARPA (intelligence). In the United Kingdom, the Advanced Research and Invention Agency (ARIA) has been launched. Other DARPA-like agencies are being created in Europe and Japan.

Nothing succeeds like success

If money is the fuel of science, ideas and preliminary data are the drill and refinery, respectively. Without ideas coupled with sufficient data to demonstrate that the ideas are sound, it is difficult, if not impossible, to find appreciable funding for science research. Funders are generally a risk-averse group with the exception of the DARPA-like agencies that understand that big wins require big plays. It is a long-dead myth that famous scientists can get funding on the basis of their name-recognition alone. Science does not allow the resting on laurels. To remain successful, scientists must continually generate good ideas for grant proposals. Do they do that alone? In most cases, no. They get help from postdocs and graduate students to make science get started and go-round: ideas → funding → data → publications (see Figure 1.1).

One can see two potential problems arising already. First, many critical steps are being performed by young scientists-in-training who might be inexperienced with both the ethics and politics of science, not to mention the nuances of the science itself. Hence, they could simultaneously be targets for exploitation and temptation. Tales abound of graduate students who are taken advantage of and not treated in such a way that their professional success is enabled. Second, each of these steps toward publication can be stumbling blocks where scientific and ethical problems might arise. Therefore, addressing potential ethical dilemmas in the context of modern scientific practices should be of some practical help. In fact, I argue here that understanding the rules of science are necessary for running a laboratory and research projects. Subsequent chapters will build upon these themes.

Summary

✓ Research science is growing increasingly complex with funding and publications becoming more and more competitive.

✓ Research ethics is a compass to find best practices in performing research.

✓ The reiterative loop of science research consists of ideas → funding → data → publications, with pitfalls surrounding each of these steps.

Chapter 2

How Honest Is Science?

> ### ABOUT THIS CHAPTER
>
> - More than ever, science is in the public eye.
> - Science is funded mainly by public sources and therefore held publicly accountable.
> - Many scientists admit to dubious and unethical behavior.
> - Older scientists appear to misbehave more often than younger scientists.
> - Unethical behavior can reach beyond activities classified as research misconduct, which is defined generally as fabrication, falsification, and plagiarism (FFP).
> - FFP carries strong sanctions.

This chapter title itself seems to be a bit odd given that the whole purpose of scientific query is to seek the truth about the world. Much like art, in which artists seek to be honest creators, science would seem to be one of those unassailable undertakings in which the pursuer has a higher calling; where idealism and truth trumps money and comfort. Ask any scientist, "Are you in it for the money?" Invariably, no scientist I've ever posed this question has answered in the affirmative even though some scientists have found the field to be quite lucrative. Contrast the PhD-level science with many other fields requiring a great deal of education. True, scientists typically don't enter science for the money, but motivations can change, and with these, behaviors can also change between the beginning and end of a career. Besides money, there are other sources of temptation in the profession. There is also the factor of sheer survival in a career choice that is populated

---

Research Ethics for Scientists: A Companion for Students, Second Edition. C. Neal Stewart, Jr.
© 2023 John Wiley & Sons Ltd. Published 2023 by John Wiley & Sons Ltd.
Companion website: www.wiley.com/go/stewart/researchethics2

with creative and smart people. Corruption can also be borne from the pressure of funding and survival gaining an unfair advantage and the "publish or perish" culture (Woolf 1997).

In addition to examining motivations of scientists, we need to take the temperature of the culture to assess how widespread is scientific misconduct. Does breadth necessarily define impact on true discovery and science itself? That is the key question. After all, perhaps science corruption is akin to cheating on taxes. Little money is at stake for a low-wage earner, but if a billionaire cheats, then significant funds are in play. Of course, there is a huge practical difference between plagiarism on an undergraduate assignment and fabricating data sets on a highly visible *Nature* publication, which would be widely read and utilized in an applied field such as medicine. But on the other hand, cheating is cheating, and little-to-none should be tolerated in science when found out.

Judge yourself

✓ Why did I decide to enter science as a profession? What were my motivations?
✓ Am I an ethical person?
✓ Am I generally tempted to cheat? In what ways?

Sanctionable research misconduct: fabrication, falsification, and plagiarism

In the United States and elsewhere, government agencies that fund scientific research have a stake in knowing if funds were spent for appropriate activities, i.e., performing honest research. Agencies, therefore, require the institutions that receive federal funding to investigate allegations of research misconduct. In most cases, research misconduct is succinctly defined as **fabrication, falsification**, and **plagiarism**, abbreviated FFP. And so, if researchers are found to be guilty of F, F, or P, then the funder and the employer of the researcher may place apply sanctions commensurate with the level of misdeed. Since the sanctions could be as dire as being denied any future funding and being fired from the job, researchers certainly have external motivations to play by the rules. Multiple offenses, very serious and egregious offenses, and causing dire harm carry heavier sanctions than lesser offenses. Fabrication and falsification are deemed to be much worse than plagiarism, but since they happen relatively infrequently, they may carry relatively low aggregate harm (Bouter et al. 2016). But the Fs carry greater sanctions than P because of the level of dishonesty and potential harm that occurs from falsifying results or simply entirely making up findings. We will see in the book some examples of great harm caused by research misconduct.

"Scientists behaving badly"

In 2002, a survey was mailed to US National Institutes of Health (NIH) grant recipients that asked them to anonymously self-report research misconduct by yes/no responses to several questions (Martinson et al. 2005). The survey was sent to mid-career scientists who had received their first full-sized (R01) grants – these were typically associate professors, having a mean age of 44 years. It was also sent to younger scientists who were supported by NIH postdoc fellowships. On average junior responders were 35-years-old. From a few thousand surveys mailed, 52% responded from the older group and 43% from the younger group. To this point, I am struck by two surprises already. First, a high proportion of scientists were willing to return the survey with the opportunity to report their misdeeds. Even if respondents could maintain anonymity, why risk self-reporting bad behavior? Confession is good for the soul, true, but why take the chance of being discovered for data fraud or other bad behaviors? Second, the sampling of scientists chosen as the study subjects was highly skewed. NIH R01 grant recipients are nearly always very well-qualified scientists who are serious about performing biomedical research. The scientists surveyed were no dilettantes. R01 grants are simply not easy to win. The 2020 proposal success rate was 21.4%. And the postdocs receiving NIH fellowships are no slouches either. These postdocs represent former graduate students whose grades, aptitudes, and research are among the best in US biomedical fields. Prior to reading the results, and based upon these initial conditions and assumptions, I would have been shocked to learn that either group reported much bad behavior; I just would not expect them to participate in research misconduct or in suspect research practices. What did the survey report?

Before we look at these particular results, the prevailing opinion, as reported by Martinson et al. (2005), was that incidence of falsification, fabrication, and plagiarism in science is probably in the 1–2% range. But in 2004 the editorial office of the *Journal of Cell Biology* (Rosser and Yamada 2004) estimated that papers containing questionable data might be as high as 20% (Anonymous 2006). One editor of a Chinese "campus journal" reported that 31% of submissions contained plagiarism (Zhang 2010). While longitudinal data on cheating doesn't exist, most people in science would agree that if we do have a 20% incidence of misconduct, or even "questionable data," then we have a huge research integrity problem with regards to fabrication and falsification. At the same time, journals routinely scan submissions for plagiarism (Butler 2010).

According to the Scientists Behaving Badly survey (Martinson et al. 2005) three researchers out of every thousand admitted to fabricating or "cooking" data. Cooking refers to altering existing data to "improve" a finding rather than outright data fabrication. The ideas of others were used (plagiarism) without proper credit for 1.4% of the respondents. For these two items, there were no differences between mid-career and early-career scientists. There were differences for at least two notable questions, with mid-career scientists being worse. Twenty-four of a thousand older, as opposed to just eight of a thousand younger, scientists used supposedly confidential information in their research. One would hope that as scientists age, they would adopt better practices and not worse. A big offense, 20.6% mid-career scientists admitted to caving to pressure from funding sources to alter their experimental design, methods, or *results* (emphasis added) as opposed to 9.5% of early-career scientists. Herein we have a striking dichotomy. Most scientists would agree that while they might be passionate about their science, that scientists ought to approach research dispassionately and objectively since the main objective of science is the pursuit of truth. However, there is a public-held assumption that funders with profit-motives or strong ideological motivations often drive or alter research results that are reported. The high numbers of researchers altering experimental design, methods, and especially results demonstrate this skeptical viewpoint of motivation has validity. I might add that the NIH is among the most benign of funders in this regard; their grantees should feel no ideological or economic pressure to find and report one result over the other. It would be interesting to perform the same survey among recipients of funds from pharmaceutical, agricultural, and chemical companies, where a research agenda is more obviously skewed toward economic impact.

If researchers are found to be guilty of many items above, there would be university and government sanctions levied against the perpetrators. But we actually hear of very few cases in which guilt is discovered and penalties are exacted. Most misbehavior in science seems to go undetected. One of the reasons why this is the case will be seen in Chapters 6 and 7. Scientists are not fond of non-anonymously reporting blatant misconduct or even sloppy science; it seems that few people want to be known as whistleblowers, or in the childhood vernacular, tattletales. This is understandable, and in the above survey, 12.5% of scientists admitted to overlooking others' flawed data or their own "iffy" interpretation of data. This figure doesn't even take into account any close examination of papers by peer-reviewers looking specifically for misconduct. Therefore, we see that many scientists would rather look the other way than "objectively" report on research honesty – even if they could do so anonymously – which is often the case in peer-reviewing of grant proposals and publications. While scientists might grumble over bad players during the social hour, they are not rushing to call out their peers, even if not doing so results in damage to science as a whole (Gunsalus 1998). It would seem that the proverbial rug covers a profundity of scientific dirt.

The Martinson study illustrates that the magnitude of incidents and self-reporting scientists with dubious behavior can hardly be considered insignificant. In fact, results in some of the categories are startling. What is even more profound is that the survey asked scientists to report on their behaviors during only the *past three years*. Furthermore, the sheer frequency of misconduct is staggering. Martinson et al. (2005) reported that best estimates of falsification, fabrication, and plagiarism prevalence were thought to be in the neighborhood of 1–2%. In their survey, however, one-third of respondents admitted to deeds that research officers would consider to be sanctionable! For early career scientists, the frequency was 28% and for mid-career scientists it was 38%!! Therefore, it is almost unimaginable to estimate how high the proportion of scientists that are guilty of gross unethical behavior over a scientific lifetime. What if they had surveyed even older scientists? If we extrapolate, well over half of researchers nearing retirement would participate in bad behavior during their final three years in science. Consider the fact that the average mid-career scientists in this survey had perhaps 20 or more years left on their career, the opportunity and propensity to participate in dubious activities is overwhelming. The survey selected relatively young scientists who had, seemingly, spent little time in science. And let's remember that the numbers reported here are likely conservative since it is doubtful that all the guilty parties would self-report their misdeeds for fear of some sort of retribution.

Do scientists behave worse with experience?

Anecdotally, I've observed that many scientists who become more savvy of the rules of science to the point that they know which ones they can break without being caught. The Martinson survey indicates this might be systematic among biomedical researchers. Of the 16 questions posed, 6 of these had statistically significantly different responses among mid-career and early-career scientists. In each case, the older scientists reported higher incidence of misconduct compared with younger scientists. The conclusion we must draw is that age and experience are important factors causing scientists to go bad. As Lutz Breitling opined in *Nature* later in 2005, younger scientists more likely have higher ideals and enthusiasm compared with older scientists who might have become jaded. "In the rough world of today's science, they are exposed to an environment in which impact factors and awards are more important than advancing the knowledge of mankind." (Breitling 2005). Sad to say, these same scientists may also be running roughshod over their trainees. He thinks that the practice of science itself, as defined by how we do modern science with its pressures and rewards, is the problem; at its root, that disillusionment is the problem. While I somewhat agree with his conclusion, I think the ultimate root of the problem must be deeper still than simply mere disillusionment. Breitling further states that he doesn't think sanctions are the answer to the problem. I'm not so sure I agree with him. In fact, another letter writer to *Nature*, Kai Wang, thinks that education along with stiffer penalties would go a long way toward improving scientific integrity. Wang, a graduate student, points out that there is little ethics education in graduate school (Wang 2005).

Judge yourself

✓ What factors of science research might cause you to cheat? For example, how do you deal with pressure?
✓ How might these be counteracted?

✓ How much of a deterrent is embarrassment and punishment?
✓ How do the results of the Martinson survey make you feel about the profession? About yourself?

Crime and punishment

Important to addressing research misconduct are reporting and sanctions. Nobody, it seems, is anxious to police science. Journal editors, to some degree, are the most proactive players in science in this regard, but clearly, peers, and especially students, are not anxious to make waves. As we'll see in Chapter 7, whistleblowing often comes at a steep price. But as Wang (2005) succinctly points out, "If the benefits of misbehaving outweigh the possibility of being punished, academic misbehaviour is probably inevitable."

Is scientific misconduct inevitable? Unfortunately, to some degree I think it is indeed unavoidable inasmuch that corruption is present in every profession; science is not immune, of course. That said, we shouldn't abandon high standards of expectation for honesty and refuse to stem the tide of unethical behavior. What can be done? First, as has been mentioned as at least a partial solution, education of the expectations and rules of science is crucial (Titus and Bosch 2010). Second, we must be aware of factors leading to potential disillusionment and corruption in mid-to-late career. Third, scientists need to self-police the profession more effectively. I am not referring to the state of mind to which Nobelist Marie Curie abhorred: "There are sadistic scientists who hurry to hunt down errors instead of establishing the truth." Quality assurance is critical in verifying that published data and information are honest and real; I think that's one duty of science and scientists. One subdiscipline in science publishing that is emerging is informatics tools to catch cheating – either pre- or post-publication. Algorithms and routines to spot plagiarism and illustration manipulation exist and should improve. Journals should be the vanguards of these activities since they arguably have the most to lose by publishing papers containing FFP or dubious results (Berlin 2009; Butler 2010).

This leads us to a discussion of consequences for being caught violating common professionally accepted standards. In contrast to many commentators, I don't think, for most violations, that the penalties should be "fatal" or completely debilitating to the scientist. Banning someone from accepting federal funds for three–five years might be an appropriate penalty for many deeds of misconduct (McCook 2009). I think that we probably underestimate and underutilize the power of professional embarrassment, or at least, the threat of embarrassment, which could be a very effective deterrent to scientific misconduct (Errami and Garner 2008; Berlin 2009; Koocher and Keith-Spiegel 2010). Clearly, the scientific community needs to debate reporting, investigating, and penalizing those who participate in research misconduct with the hope and expectation of rehabilitation.

What is the role of institutions that uncover researchers' misconduct? Should they go public with announcements that their "famous scientist X" has been found responsible for committing research misconduct? In most cases, it appears that all aspects of misconduct cases are kept quiet by university administrators. Few want to call attention to trouble at their institutions. In the past 30 years, I can recall a small handful of press releases by universities to that end. It seems to me that there could be some value in demonstrating that institutions are vigilant to prosecute wrongdoing and declare that they've performed fair investigations to protect the integrity of their research programs. Perhaps administrators are counseled that there is potentially more to lose than gain in bringing bad news public. Which is to say that they might believe they would bring more embarrassment onto themselves than would their former employees who were found responsible for research misconduct.

A prominent moment that usually carries professional embarrassment is when a paper is retracted, especially from a high-profile journal, and especially if it is retracted because of scientific misconduct. Indeed, retractions continue to trend upward (Figure 2.1). Of course, papers may be retracted by the authors when an honest mistake is made; these are understandable and not sanctionable. Those papers retracted because FFP are in another league. In a landmark case, *Science* retracted, without the authors' permission, a paper that contained fabricated data (Normile 2009). While it took the journal four years to arrive at this action, and after they tried and failed to contact all authors to obtain their permission for retraction, it still took gumption to take the action of pulling the paper from publication. Papers are usually retracted as a last consequence when it is found that data are so wrong that the paper's main conclusion is no longer

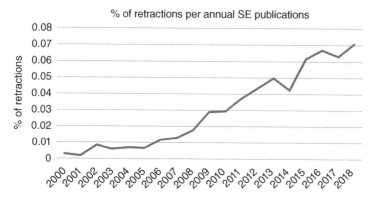

Figure 2.1 Retractions of a given year's publications as a percentage of papers published in science and engineering (SE).

Source: The data are from the Retraction Watch Database that include NSF data. The figure is reproduced here by permission of Ivan Oransky/Retraction Watch.

justified. In this particular case, some of the authors of a 2005 paper agreed that the paper on tracking proteins should be retracted because data were fabricated. The method to track proteins that was presented called "MAGIC" is still valid and should be patentable according to the senior author of the paper, even though the published results and data were indeed fabricated. Muddying the water even further in this case is that a company was started by the same author and others to commercialize the method. The waters here appear very muddy indeed. However, one thing is certain, the authors, especially the senior authors, are paying dearly for their misdeeds. The institution where the researchers were employed, the Korean Advanced Institute of Science and Technology (KAIST) in Daejeon, South Korea, has taken the dramatic action of dismissing lead author Tae Kook Kim. Furthermore, his company sued him for fraud. In response KAIST developed educational programs to ward off trouble and to "promote research ethics and integrity." (Normile 2009). Good idea!

Should misconduct always be punished so harshly? A case can be made that leniency is often appropriate and that harsh sanctions are unfair. In the United States, the Office of Research Integrity (ORI) is the investigative and sanctioning body for research misconduct. The charges they investigate fall under the heading of fabrication, falsification and plagiarism (FFP). If found guilty (or "responsible"), researchers end up listed on the ORI "Administrative Actions" website and these people are banned from obtaining federal funding (debarred), typically for three–five years. It used to be that they kept a lifetime list of cases, but currently the ORI lists only active debarred cases. This makes a lot of sense. Allison McCook (2009) discussed the reasons why the old system could be harsh given that once the information hit the internet, it never disappeared. She spoke with several formerly listed investigators who were found guilty of relatively minor research infractions. These researchers claimed that their misdeeds have resulted in their being forever stigmatized. One example was an assistant professor at a large public university who included data in a grant proposal that he really didn't have authority to use. His story was that he prepared a grant proposal using preliminary data from a collaborator at a startup company – one that he'd recently worked with on another project – who was to be a co-investigator on the proposal in question. A few days before the proposal was to be submitted, the collaborator informed him that the company didn't wish to participate. At that point the assistant professor should have removed the data in question but forgot to do so, even though he did remove the collaborator from the proposal. The company found out that their data were included in the proposal and complained to the assistant professor and then the university, which was then required to forward their findings to the ORI. The assistant professor admitted that he was guilty of research misconduct as charged. Although his story ends relatively happily for the researcher (he received tenure a year after he signed papers admitting guilt and is a productive scientist), other researchers struggle with the stigma of similar-magnitude misconduct. McCook (2009) chronicles researchers whose stories ended with prolonged ramifications – not so much by the ORI sanctions, but because the sanctions are forever searchable on the internet. Like naughty pictures of sophomoric high jinks posted on Facebook that come back to haunt job applicants and politicians alike, research indiscretions have long-lasting consequences. One researcher has had a difficult time finding research employment despite a rich publication record. Out of desperation he changed his publication name from his real name to escape the stigma. Another researcher left academic science, experienced stress-related health problems, and then later started a small lab-based company in his kitchen. I believe that scientists who have made mistakes and then learn from their mistakes, should, in many cases be allowed to resume their careers.

Judge yourself

✓ How do you feel about "crime and punishment"?
✓ What can be done to create a more accountable science environment in your lab?

So, on second thought, I'm not convinced that perpetual embarrassment is the solution to research misconduct, especially for relatively minor offenses. I think that education on best practices can be the most helpful intervention for young researchers (Titus and Bosch 2010). Slowing down to think about ramifications of research and its representations are also useful. Certainly, the punishment should fit the crime and experience and history should be considered. Do we live in such a corrupt science culture that we can't trust most research outcomes? I think the answer is "no." In spite of reports of high levels of misconduct, I believe that most published science is sound. None of the McCook cases mentioned above involved retracted papers. Often a discerning eye finds data that "doesn't look right" or results that are too good to be true, and it does not hold the credibility as better papers. Between discernment and replication of experiments, science truly is a self-correcting enterprise, at least most the time.

This chapter focused on making bad ethical decisions and not incompetence. Honest scientists make all kinds of judgments about which experiments to perform, what data to collect and report, and what data should remain invisible to the public. These decisions are not relevant to misconduct but rather are components of professional judgment. Non-scientist Napoleon Bonaparte teaches us well, "Never ascribe to malice that which can adequately be explained by incompetence." There are as many levels of competence and shades of gray as there are numbers of scientists. Anyone who works in biology, for example, knows that there is great variation in biological systems and that some outliers don't represent the reality of the research accomplished. Sometimes, simply bad science (incompetence) is the subject of retraction of publications instead of misconduct. Indeed, when I review papers, I typically look for honest mistakes in experimental design, execution of experiments, and data interpretation. That is really the purpose of peer-review. Or as Napoleon Bonaparte said, "Never ascribe to malice that which can be adequately be explained by incompetence." Misconduct, as defined by FFP, requires the intent to deceive to be "proven" (Shamoo and Resnik 2003). Indeed, intent to deceive is often hard to prove.

Joseph Macrina (2005) discusses the question, "Who needs ethics?" (Weston 2002) in his book and frames the answer in light of the science profession's needs themselves. It is obvious to me that the needs are greater than we could've imagined. The readers of this book will delve into the rules of science and learn how ethics guides the brightest path while building a successful research career. We won't worry so much about falling into the hole of FFP, but how ethics can help us stay on high and solid ground. After all, most of the research ethics decisions that scientists make are not "what data will I fabricate today?" Or, "which paper would be the best one to plagiarise?" But rather scientists must most often judge their own actions to check for objectivity and pure honesty in doing science.

Case study: the case of a pain in the lab and the question of sanctions

Professor Swarmy Ratticus, who was a researcher at a prominent West Coast medical school, ran a lab of researchers who specialized in pain research. He was known as an expert in the field, since he'd been on faculty for 20 years, currently as a full professor. He was a co-author on over 80 research papers. For his research Dr. Ratticus uses laboratory animals. In US universities, the use of animals in research is regulated at the university level by an Institutional Animal Care and Use Committee (IACUC). The IACUC requires laboratory protocols that minimize an animal's pain and suffering and they do periodic lab inspections to make sure the rules are being followed. Dr. Ratticus and his team of postdocs, Drs. Lamb, Hogg, and Hens were not always a happy bunch. It seems that Dr. Ratticus would pit postdocs against each and told them that the IUCUC committee could not be trusted. "We're the experts, not them," he would declare and told the postdocs to hide any sick animals just before IUCUC inspections. In addition, Dr. Ratticus would sometimes remove data points to make figures that were to be published appear to be statistically significant. This happened to

Dr. Lamb and Dr. Hogg on separate occasions. Because they were intimidated by Dr. Ratticus, they didn't do anything about the figures and their papers were published with the questionable data. Dr. Hens was a relatively new postdoc and had not yet been a co-author with Ratticus. But all three postdocs had been subjected to bullying and coercion during one-on one meetings with Ratticus. During a late-night meeting among the postdocs, they shared stories about the coercion, and possible research misconduct, and then also discussed the hiding of sick animals from the IUCUC. At that time, they decided to report Dr. Ratticus to the IUCUC, which led to an investigation by that committee as well as a second investigation by the university's research misconduct committee. The latter committee was also forced to investigate Drs. Lamb and Hogg for research misconduct since they knew about the research misconduct in the papers they authored but allowed them to be published anyway. They all were found to be responsible for falsification of results. They determined that Dr. Ratticus was "intentional" in the misconduct, meaning he was purposeful in the falsification, which is the highest degree of intent. The postdocs were found to be "reckless" (the lowest degree of intent), but the committee also wondered if they were "knowing" (the middle degree of intent). During the misconduct investigation's discussions, they discussed the possibility of imposing more dire sanctions for Dr. Ratticus than the postdocs.

Discussion questions

1. Do you think that everyone should be punished the same, or do you think there should be different punishments rendered?

2. How much weight should the committee place on differences of experience among the three researchers?

3. What about the power differential among people in the lab?

4. Could have there been a better way to report Dr. Ratticus' bad behavior compared with what the postdocs chose?

Summary

✓ Research misconduct and unethical research activities are extremely common with one-third of scientists self-reporting offenses within a three-year period.
✓ Scientists, reviewers, editors, and publishers are responsible for ensuring accurate scientific information is reported.
✓ Sanctions include disbarment from funds and public and professional embarrassment.

Chapter 3

Research Misconduct: Plagiarize and Perish

ABOUT THIS CHAPTER

- Plagiarism, defined as using others' ideas, sentences, or phrases without citation, and is, by far, the most common form of research misconduct.
- Plagiarism is easily avoided.
- Plagiarism is becoming more easily detected thanks to computational and networking tools.
- Some self-plagiarism is typically unacceptable in instances where the source material has been published.
- Recycling your own writing is usually acceptable when it has been previously unpublished.

Plagiarism is claiming others' ideas, sentences, or phrases as one's own. In professional writing, plagiarism is typically considered fraudulent and intentional by the fabrication, falsification, plagiarism (FFP) definition. In student writing, plagiarism might be done without any malintent. For example, students with a lack of command of language and science also could be tempted to plagiarize as a survival mechanism without understanding the rules and ramifications. Nonetheless, in science, fairness and honesty dictate that others' work is recognized and cited in scientific literature, grant proposals, and coursework. This is the widely accepted standard among all academics; scientists or otherwise. While it is common, plagiarism cannot be tolerated or condoned.

Research Ethics for Scientists: A Companion for Students, Second Edition. C. Neal Stewart, Jr.
© 2023 John Wiley & Sons Ltd. Published 2023 by John Wiley & Sons Ltd.
Companion website: www.wiley.com/go/stewart/researchethics2

To me, plagiarism is the most boring of ethical offenses; largely borne of both academic laziness and ignorance, and being contented to remain in this state. Or, in haiku:

Cannot synthesize?

Lazy about your writing?

Why not plagiarize?

Shamoo and Resnik (2003, p. 50) refer to authorship as "perhaps the most important reward in research." But then follow to declare that "publish or perish … is a grim reality of academic life." Unlike the situation when we were young where the teacher assigned a dull theme paper topic, research scientists get to write and share their hard-earned results with the world by writing journal articles. Thus, I agree with Shamoo's and Resnick's first statement about writing being rewarding, but I admit that I can't grasp how enthusiastic scientists could ever view publishing as "grim." Therefore, my view of plagiarism is when authors feel like they must fake the acts of creativity and communication. Indeed, communicating results is vital to scientific research (Macrina 2005). Without effective communication, scientific results cannot be understood.

Plagiarism is an umbrella term to mostly describe a wide range of co-opting text, even one's own (self-plagiarism). Plagiarism ranges from mild-to-extreme, and I think that the extreme cases are typically the only ones uncovered. Mild plagiarism is seen, I think, oftentimes as a legitimate shortcut that is often tolerated. There are various reasons for this. First, no one really wants to police science writing, and so only blatant plagiarism is reported and punished. And even then, sometimes it is not reported. Second, some suspected or alleged plagiarism may not actually be plagiarism. Third, some "plagiarism" is allowed, such as some self-plagiarism – let's call it "text recycling" instead – is acceptable and even useful in science. No matter what your view on plagiarism, there are electronic tools that are being increasingly used to discover it in professional and student writing, and so reporting and punishment will become more important in the future. These developments and predicaments will be examined in more detail later.

There does seem to be various cultural views about plagiarism, but it is clear that in science, there exist international standards for "owning" one's writing in which plagiarism is not allowed. Mimicry in nature often confers an evolutionary advantage, but in scientific writing, it can be a career killer. In some cultures, imitation in writing, i.e., plagiarism, might be construed and accepted as a form of flattery, but in research and coursework at universities, it can lead to a student's dismissal. A professional but plagiarizing scientist can be subject to censure by employers, journals, professional societies, and granting agencies. This chapter will examine what plagiarism is and what it is not, and it will examine gray areas in scientific literature (e.g., using the author's own writing as source material) where there seems to be a fine line between right and wrong.

Plagiarism abounds. Errami and Garner (2008) estimated that perhaps 200,000 of the 17 million papers indexed in Medline, a database of life science papers, might be duplicates. Some of the 200,000 are outright plagiarism, but perhaps 30 times more are simply duplicated articles by the same author. Is this an allowable practice? This chapter will seek to define plagiarism in order to, hopefully, avoid it altogether, and definitely point us toward best practices in professional science writing.

Ideas

Surprisingly, there are some scientists who seem to be devoid of many original ideas and will not or cannot generate ideas. This is a huge problem that might be insurmountable when attempting to sustain a career in research. Good science is epitomized as the implementation of novel ideas and sound methods to address pertinent questions and problems. Indeed, these, to a large degree, define what is fundable via grants and contracts. After all, who wants to fund something that is passe and not novel? And, without money, there will be little scientific research. But ideas and lines of scientific research are continually recycled, aren't they? To some extent this is true, but there should always be an increment of novel contribution to warrant publication in the scientific literature (for the most part anyway, as we'll see later). Perhaps more than an increment is needed to win grants. Certainly, if an idea has been proffered by someone else and you mention it in a paper, the source should be cited. This gives credit where credit is due and also impresses your peers that you know the literature. No one likes someone who pretends to have thought of everything first. Science is about ideas and data building on existing ideas and data – Francis Newton's notion of him standing on great shoulders coming before him.

Sentences

Copying and pasting sentences from one published paper to another is plagiarism. The surest way to get caught is to also include internal referencing from the source paper. For example, Paper A contains the following sentence:

> What distinguishes GFP from other reporter genes is its ability to fluoresce without added substrate, enzyme, or cofactor (Prasher et al. 1992).

Now if we see this particular sentence verbatim in Paper B, citing Prasher et al. (1992) and without citing Paper A, then plagiarism has occurred. This is an open and shut case. Even if citing Paper A, the author of Paper B would need to put the above sentence in quotation marks, and, in practice, this is almost never done. The author of Paper B would be attempting to convince us that he knew this fact and was clever enough to write this sentence. People often get caught plagiarizing if their own writing is visibly weak and an author includes plagiarized sentences of a vastly different style and quality. I've had editorial duties for several journals for over a decade and have seen manuscripts submitted in which entire paper abstracts are lifted verbatim or with a few word changes. Some people seem to love the copy and paste style of writing.

Phrases

Copying or the apparent copying of phrases is not as clear cut as the copying of sentences. That is because certain phrases are optimal expressions of ideas or descriptions, and so we're not surprised to see common phrases reoccurring in the literature. People get suspicious about plagiarism if familiar phrases get used quite frequently in a paper, especially if they are from the same apparent source. If possible, specialized phraseology should be cited, especially if this denotes a unique contribution from someone else. This is a gray zone, which calls for judgment about specialized knowledge and phrases.

A hoppy example

In the same issue of the regional specialist journal *Southeastern Naturalist*, there appeared two unrelated papers on swamp rabbits. The authors between the two papers lived and worked in different US states.

Fowler and Kissell (2007) begin the introduction of their paper saying, "*Sylvilagus aquaticus* Bachman (swamp rabbit) is primarily found in bottomland hardwood forests along rivers, streams and swamps (Allen 1985; Chapman et al. 1982; Kjolhaug and Woolf 1988)."

In the second paper in the same issue, Watland et al. (2007) start their paper with, "*Sylvilagus aquaticus* Bachman (swamp rabbit) is a representative species of bottomland hardwood forests, occurring primarily in swamps, river bottoms and lowland areas (Chapman and Feldhamer 1981)."

In each of these papers, I read the cited references to answer two questions. First, was there any plagiarism? And second, were the references appropriate to cite? To answer the first question, in my view there was no plagiarism. Each topic sentence by Fowler and Kissell (2007) and Watland et al. (2007) were distinct from those in their references and likely just happened to converge on similar wording by chance. This example teaches us that happenstance is a kinder and more accurate explanation than plagiarism in many instances. To answer the second question, there were several general papers written about swamp rabbits describing their habitat and habits during the 1980s. Each that I surveyed had seemingly appropriate features to make them citable in this context. It could be argued that the fact of swamp rabbits living in the swamp could be obvious and general knowledge, but the authors of both papers played it safe in citing appropriate references. If the authors were guilty of anything, it was trying to impress their peers that they knew about these publications that stated that swamp rabbits live in swamps.

What is plagiarism, really?

So, aside from these textbook definitions, what is plagiarism, really? When do people get concerned that a paper contains plagiarism? How much originality is needed to define a work as original, and how much similarity can be tolerated before it is declared plagiarism? Let's start with ideas. Very little in science is totally novel and plagiarism of ideas is the hardest to prove. By its nature, most science is derivative, but it is important to cite appropriate literature from which a paper might be derived. Typically, if ideas are plagiarized, then sentences and phrases tend to be plagiarized too, and so this is a good place to start in the checking for plagiarism.

Judge yourself

✓ What is your ethical background with regards to plagiarism?
✓ Were you taught about plagiarism in school?
✓ How good are your communication skills and command of the languages that are commonly used in science, especially English?
✓ Are you comfortable and confident in writing?
✓ Do you think you have good ideas?
✓ When you write, do you feel like you have something to say?
✓ What things can you do to improve your science communication and writing?

Indeed, it is rather easy to find apparent plagiarism if one wishes to look hard enough. Most scientists have better things to do. But, if we spot a phrase or sentence that sounds familiar, one of the easiest ways is to enter it into Google or some other search engine within quotation marks and analyze the results. Let's look at an example. I wrote the first sentence in this chapter (the definition of plagiarism) before I read any papers or did any internet searches on plagiarism. The definition I offered was one that was burned into my memory from my Chief Justice days in the early 1990s. So, I did a Google search of: "Plagiarism is claiming others' ideas, sentences, or phrases as one's own." I was genuinely surprised to see that this sentence in quotations yielded no results. So, I tried smaller phrases until the transposed "ideas, phrases, or sentences" finally turned up one hit. Additional transposing of the words returned other Google hits. So, is my original sentence plagiarism? No, because I used my own knowledge and memory of the subject. Alternatively, I could have simply looked up a definition and quoted and cited it. Also in my defense, I also had no intent to deceive. At its core, plagiarism is a misconduct characterized by misrepresentation. Also in my defense, one could argue that the general definition of plagiarism falls under the category of common knowledge, and so it would be difficult to prove plagiarism in such a case. For example, it is not plagiarism to write, "The sky is blue." That is the case, even though this sentence has certainly been written in thousands of works. In science, there is any number of gray cases that could fall either way, but the general expectation is that the author will provide new knowledge somewhere in a paper; in best cases, a lot of new knowledge. Fields of science are composed of knowledge that range from "something every good physicist knows," for example, to obscure and arcane knowledge that should always be cited.

How many consecutive identical and uncited words constitute plagiarism?

This is the question many people ask and I don't think there is a clear answer, although other people obviously disagree with me. The University of Idaho Information Literacy module on plagiarism states that the use of others' two or more words in a row should be placed within quotation marks (http://www.webs.uidaho.edu/info_literacy/modules/module6/6_4.htm). I find "that criterion" to be untenable, "albeit conservative" to catch all "possible plagiarism." This system would result in distracting quotation marks littered throughout the document. So, I think two is too few. I've seen other anti-plagiarism enthusiasts make the case for three-, four-, five-, and more words in a row to be problematic. I recall seeing a company that advertised their writing service that included translating a text written in one language into English for publication. One "side-product" they sold was assuring that the publication was plagiarism-free as defined by making sure that authors had not copied six consecutive words. To me it matters more which identical and consecutive words are found. I believe that the focus on numbers of consecutive copied words to be futile. A better focus, in my opinion, is for authors to develop effective writing skills so that they become competent at original writing. When authors want to use the perfect sentence or phrase that was coined by another writer, then they should put it in quotes to be safe. In most cases, authors should simply put phrases and sentences in their own words and provide an appropriate citation. Probably the best thing that authors (scientists) can do is to make sure they have something interesting and novel to say in the first place. Having an exciting message is fuel for great writing.

Self-plagiarism and recycling

Plagiarism is using others ideas, phrases, and sentences without proper attribution or citation. **Not allowable.**

Self-plagiarism is plagiarizing your own published work; typically large amounts of copied and pasted segments. **Generally not allowable.**

Recycling is copying and pasting your own writing from an unpublished source to another document that might be published or unpublished. **Generally allowable.**

Self-plagiarism is simply using your own words in more than one document; typically the source document has been published before, hence the plagiarism (negative) label. There is a wide variety of views about self-plagiarism, ranging from "duplicative publication is unethical because it is wasteful and deceitful" (Shamoo and Resnik 2003, p. 83) to the notion of the impossibility to steal (plagiarize) from yourself. The safest haven in writing and best practice is to equate self-plagiarism as being equivalent to plagiarism (Green 2005; Berlin 2009). There is a certainly a functional difference between recycling your own writing when it has been unpublished previously (not plagiarized) and self-plagiarizing from a published paper that will be used in another published paper (double publication = plagiarism). Recycling is allowable and is not self-plagiarism. The functional scalpel here divides between published and unpublished work, where publishing the same information more than once is unethical. These days, self-plagiarism is viewed at the same level as plagiarizing another author: not allowed.

I think that a common-sense approach about sources and sinks is helpful. For recyclable sources, such as your own grant proposals, notes to yourself, and unpublished white papers, the key is that they are not published. But as sources move from non-published to published, the verbatim copying should be avoided. For sink documents, i.e., the document that work is being pasted into, there is more leeway for allowance for recycling without ethical ramifications. The goal is for each published document to be unique and containing new and valuable information. Whole manuscript duplication should be avoided, unless it is simply a reprinted manuscript and indicated as such in the front matter. Copying and pasting among certain documents is completely allowable as long as the source is unpublished.

Judge yourself

✓ How do you feel about copying and pasting?
✓ How do you feel when copying and pasting?
✓ What is your view about self-plagiarism?
✓ Do you rely on much copying and pasting during writing a project?

Path of science publications

Recycling of most conference writing is allowable

After an idea is hatched and data are collected, the consummate publication, the one that "counts" in the eyes of other scientists and the people evaluating your career is the peer-reviewed paper published in a journal. Spend your time doing your best writing here. But many (most?) times a peer-reviewed publication is born as an abstract for a talk or poster at a scientific meeting. For a talk, the abstract is the only writing that might represent a project to the public. And in most cases, the abstract will only be available to people who paid to attend the scientific meeting and not generally accessible to others. The same is true for a conference poster. A poster might contain the draft writing that will morph into a scientific paper that will someday be submitted for publication. The same is true for a thesis or dissertation. Therefore, your own abstracts, poster text, and dissertations can be completely recycled into the manuscript slated for peer review and subsequent publication in a scholarly journal. That is, feel free to pull text from any of your conference abstracts and posters to help write your own peer-reviewed publications. A smart scientist presents a poster that can simply be flipped into a publication with a little tweaking. If that resulting publication is a peer-reviewed publication, then the self-plagiarism must then come to a halt for that particular writing stream. The peer-reviewed publication is the dam. Conference proceedings articles are often considered as writing "halfway houses," so they are also fair play for self-plagiarism (um, recycling). At least that's the case in many fields. Why are these allowable sources? Conferences and conference proceedings are considered to be appropriate for vetting ideas and results that are perhaps not quite mature or complete. In the life sciences, conference abstracts and proceedings are not the finished products, but rather places for projects to receive their first rounds of criticism so that they can be refined for the peer-reviewed publication that comes later. That's okay. Ever since the days that typewriters were shelved for word processors, no one writes everything from scratch, and it is important to determine what documents are worthy to build upon and which ones should be torn down and buried. You must make wise judgments for yourself. And you are only allowed to build upon your own abstract or poster; not someone else's. In certain fields, such as computer science and engineering, many conference proceedings are peer-reviewed and constitute the final publication in the stream. This almost never happens in my own field of biology. Therefore, best practices vary among fields of study and you have to know the particular rules and best practices in your particular field. It is therefore not surprising that the software package, SPLaT (Collberg et al. 2003), for the detection of self-plagiarism, was conceived as a tool to help prevent self-plagiarism in computer science. The reasoning behind SPLaT and the argument against self-plagiarism is nicely stated on the SPLaT homepage (http://splat. cs.arizona.edu). "It is our belief that self-plagiarism is detrimental to scientific progress and bad for our academic community. Flooding conferences and journals with near-identical papers makes searching for information relevant to a particular topic harder than it has to be. It also rewards those authors who are able to break down their results into overlapping *least-publishable-units* over those who publish each result only once. Finally, whenever a self-plagiarised paper is allowed to be published, another, more deserving paper, is not."

Patents

After scientists gain more experience, they know a good publication or good publication nugget when they see one; a paper that leads to a noticeably substantial contribution to science. A publication idea might not be given first at a conference. In fact, a really hot paper might be presented at a conference only after the publication has been submitted because of patent issues. For patentable inventions, it is best practice not to publicly disclose information – "prior art" in patent lingo – before a patent is filed. A publication could very well enable someone else to reduce to practice your invention. Of course, this is the whole point of scientific publications: to disseminate knowledge so that it can be beneficially used by others. Filing for patent before the knowledge is disseminated protects an inventor's commercial rights by protecting patentability. It is perfectly acceptable also for a patent application to have self-plagiarized material – even though a patent attorney is writing the patent application on the inventor's behalf.

Review papers and book chapters

Again, as a scientist develops a reputation as an expert in the field, there will arise numerous invitations to author book chapters and review papers. These are often good opportunities to contribute to the foundations of science. Review papers published in journals are also peer reviewed and these count toward a scientist's dossier of peer-reviewed publications; sometimes in its own category. Sometimes the same is true for book chapters, but these are not created equally. Certain books, it seems, are published by obscure publishers and don't seem to receive wide readership or citations. Other books seem to be taken much more seriously as citable source material. For our purposes here, we won't judge the prestige of the chapter or the book and how much it will be read and cited. Rather, we will attempt to develop best practice guidelines about self-plagiarism and writing recycling guidelines for these documents as sources or sinks.

Both review papers and book chapters are opportunities for an author to synthesize old information into a new publication in which existing knowledge in the literature is coalesced into summaries and opinions about a field of research and where it might be headed. A truly well-written review paper has value in that it synthesizes new insights from previously published papers. Bad review papers read like shopping lists of who did what, but without the synthesis. Reading review papers are certainly a recommended best practice for young scientists trying to learn the broad landscape about a subject: the platform to launch new science within a context of what's already known.

Since journal-published reviews are peer-reviewed (but not all book chapters), they are also entered into the canon of the scientific literature. Inevitably, after a good review paper is written, someone editing a scientific research book will ask for that same information to then be reworked for inclusion in said book. Certainly, it acceptable to do this, as long as self-plagiarism is avoided. For example, whenever I'm asked to author or coauthor a book chapter, it is nice to start from one of my prior review papers or a grant proposal as a skeleton document about what is known in the field at the time, which can then be updated. The expectation from most book editors is that a skeleton document is a good place to start, but nobody wants to see a reproduction from source to sink. It is not acceptable to copy and paste the latest review paper, change the title, rework the referencing format, and call it a new contribution. Typically, my colleagues and I will do extensive editing to rewrite sentences, change the flow of an existing paper by altering the emphasis and headings, and update the information and references so that the new contribution has value on its own merit. We want the next publication to be better and more informative than the last one we wrote. Many of the ideas may have some resemblance of those found in the prior work. That's fine, since the ideas included in the review paper are typically the ones a book editor wants included in the book in the first place. But care should be taken not to simply reproduce another work. Furthermore, there should be no intent to deceive the reader. Where appropriate in the book chapter (sink), it is important to cite the review (source) to alert the reader of the prior published paper. See the self-plagiarism/recycling network (Figure 3.1).

Grant proposals

In several instances in this chapter, grant proposals have been mentioned. Writing grant proposals to support the funding of research is a necessity in modern science. A government agency will issue a call for proposals, and scientists respond with their ideas and plans about how to address the call with their proposals. While it is not ok to plagiarize another author in your proposal, recycling your own writing is perfectly acceptable when writing grant proposals. I've recycled any number of writings into grant proposals, and that is useful to express the idea of the proposed research in a time-economic fashion. Grant proposals are not publications. In fact, they are considered to be privileged documents for review purposes only (see Chapter 9).

Recycling network

Original building blocks (Must be fresh)	Recycling candidates	Must be fresh
Ideas	Conference abstract ⟹ Cloned abstracts Posters	Peer-reviewed journal articles
Early data ⟹		Review papers
Grant proposals ⟹	↘ Conference proceedings ↗	↙
Extensive data		Book chapters
Intellectual property (patents and trademarks)	Grant proposals	Articles for trade and popular magazines
	-INTEGRITY-	Books

Figure 3.1   A writing recycling network scheme. This scheme indicates some likely sources of text that can ethically be reproduced (recycled) without being deemed plagiarized in the publication sinks as shown by arrows by the same authors. The final products of research (toward the rightmost part of the figure) should seldom be recycled by authors to avoid self-plagiarism. Care must be taken to cite sources and to make sure that credit is given where due.

New writing certainly goes into grant proposals, but that can be legitimately copied and pasted into posters and publications alike or other grant proposals written by you. Once the peer-reviewed publication is written, it can be recycled into additional grant proposals, but not into new publications. The idea behind grant proposals is to not disseminate new knowledge, but to propose an idea and project for funding. Therefore, your own recycling of text describing methods sections, literature reviews, rationales, and justifications is often useful when writing proposals. The ethical dilemma that we'll see later is that it has to be for work that is nonoverlapping. That is, one could envisage using the same methods and techniques to address a new problem and project. Grant proposals are special cases in the world of writing. Even though proposals are not publications, per se, it is never acceptable to fabricate data or plagiarize others in proposals. Nor is it acceptable to plagiarize ideas. These would all be unethical and sanctionable. To that end, some granting agencies use text-similarity (plagiarism)-checking software on submitted grant proposals.

Writing for coursework

Since grant proposals are not real publications and are available as sources or sinks for self-plagiarism, it may make sense to people that papers written for high school, college, and university coursework could fall under the same category and be treated under the same rules as grant proposals. If that was the case, then why do professors get upset when students plagiarize or self-plagiarize? The reason is, even though themes and term papers are not real publications, the expectation from all professors is that students will produce something new and related to specific elements in the course. The big idea is that performing literature searches and reviews, synthesis, and the process of writing new material for assignments comprise a substantial act of learning. Therefore, it is not allowable for you to self-plagiarize your own writing for courses – no learning or real writing takes place when that occurs. Cheating in this way defeats the whole purpose of learning to write and synthesize ideas for student development. It is not a shock why some colleges and universities permanently dismiss students who are found to be guilty of plagiarism.

Self-plagiarism and recycling redux

To dissect the differences why plagiarism and self-plagiarism is unethical and why certain text recycling is ethical, we need to revisit the previous chapter on the foundations of ethics. One benchmark of ethical decision-making is based on minimizing harm and maximizing benefit. For plagiarism, the person whose ideas, sentences, and phrases are stolen has been violated and the thief takes the credit for the presented scholarship. Both people are harmed. To test whether individual cases of copying and pasting is self-plagiarism (unethical) or recycling (ethical), we need to investigate potential harm. If the reader expects a totally new synthesis, such as the case of a review article in a journal, then no self-plagiarism is acceptable. Since edited books are often collections of previous ideas that are compiled into a single volume, there is more lenience for recycling, but self-plagiarism is not allowable. If a scientist were to publish multiple research papers that are self-plagiarized and contain identical or near-identical data. This is known as "salami slicing" – parsing up what could have been better presented in one paper into two or more papers – then science and society is harmed, since it is assumed by the scientific community that each paper should not be repetitive or others (see the box on salami-slicing). Each case can be analyzed in this way to identify sources and magnitude of harm to judge what the best actions should be. With little practice, most scientists easily discern best practices in their fields.

Salami slicing

Salami slicing is the practice of double publishing or publishing several papers from a study instead of fewer larger and more comprehensive papers. Salami slicing, also known as the production of the "least publishable unit" could lead to repeating (self-plagiarizing) methods, data, and introductory materials, and unnecessarily dividing up datasets; all of which are unethical (or if not unethical, certainly not the best practice). Salami slicing is not in the best interest of science as the result is convolution and confusion of scientific results. If a unified dataset is divided into salami sliced papers, it defies logic and expectations among the community of scientists. The best practice is to publish papers in "natural" units, whether they be small or large papers. I've written published techniques papers that are very short. I've also written rather long papers with lots of results that fit together in one unified story. Sometimes it is better to imbed a new technique within the context of a bigger paper and sometimes not. It is important to consider the ethics and readability of parsing up a study rather than keeping it intact if it would better serve science to keep all the subtopics together as a unit. What would the reader rather read: several smaller papers (or a subset of these) or one bigger and more comprehensive paper? Which is more useful? One growing problem in science is the "avalanche of low quality research" (Bauerlein et al. 2010). This avalanche refers to papers that are written but never cited – more than half of scientific papers are cited within their first five years of publications. Let's let ethics be our guide – why publish a paper that no one reads or cites?

Judge yourself

✓ How would you feel about someone plagiarizing your work or stealing your ideas?
✓ How much do you want to trust authors and their papers?
✓ How would you feel if you discovered an author had published duplicate (or near-duplicate) papers on the same material in two different peer-reviewed journals?

Dissection of a manuscript for plagiarism

A scientific manuscript is typically composed of five sections: (1) abstract, (2) introduction, (3) materials and methods, (4) results, and (5) discussion. Tables, figures, and references are also included as well as a short acknowledgments section. The abstract is typically the last part of a paper to be written, so it should always be freshly synthesized to fit the content and tone. The introduction section might contain ideas and progressions that are common to many papers on similar subjects. For example, I used a derivation of a sentence published in 1996 in a multiauthored 1997 paper. Both of these papers were about applications using the green fluorescent protein (GFP) in trans-genic plants. GFP had only been used in a few studies in plants, and only beginning in 1995.

My 1996 paper: "What distinguishes GFP from other reporter genes is its ability to fluoresce without added substrate, enzyme, or cofactor (Prasher et al. 1992)."

My 1997 paper: "GFP is the only well-characterized example of a protein that displays strong, visible fluorescence without any additional substrates or co-factors (Heim and Tsien 1996)."

Note that the idea is the same in both sentences, but that the second sentence is constructed better and says more. Note also that the reference was updated to reflect the new scientific development. In both papers, the other novel properties were discussed in the same order. This is not plagiarism. Even if the sentence from the 1996 paper was not mine, the 1997 paper would not have been a case of plagiarism.

Like the introduction section of a paper, the descriptions found in the materials and methods might also be similar to other papers by the same or even different papers. Chemical names, formulae, instrumentation, organisms, geographic locations, etc., might be quite commonly used among closely related papers or papers using similar methods that authors should describe. One might argue that citing a paper in which these items were perhaps first listed could be more appropriate than a verbatim listing. If it is a common scientific procedure that has a memorable label and is at least somewhat routine, then this practice is entirely appropriate. But since the purpose in science is to communicate clearly about what was done and learned, then oftentimes, lists and descriptions are necessary to include as a service to the reader. Again, think of the Golden Rule. What practices do you want to be used in papers that you read? What practices increase clarity? Indeed, there is often only a few logical ways to phrase certain statements and these might be repeated from time to time. This is typically not construed as plagiarism. But don't copy and paste methods sections from one of your (or others'!) papers. Certainly, the safest route is to cite the best paper(s) for a method and then offer any updates or modifications briefly in the new paper.

Results and discussion sections should always be written fresh and special care should be taken not to plagiarize. The discussion section is always the most difficult to write since the author is putting the new results in the frame of reference of existing published knowledge. Typically, if there are sentences that seem familiar in a discussion, then it warrants investigation of possible plagiarism. Since it is such a tough task to write a good discussion section, this is a logical place for the lazy plagiarist to plagiarize. This is a shame because the discussion section is best opportunity for real creative expression in a scientific paper.

So as we see, similarities among ideas, sentences, and phrases must be placed in context of each paper and the sections of each paper. Similarities are not created equal.

Tools to discover plagiarism

There are a number of anti-plagiarism services and tools on the market. The following websites that describe for-fee services:

www.turnitin.com

www.ithenticate.com

www.plagiarismscanner.com

www.plagiarism-detector.com

www.plagiarismdetection.org

One can also use Google or other search engines to find text that is questionable. There is no doubt that publicly available scripts will be written to help catch plagiarism.

iThenticate

iThenticate is a for-fee service first offered in 2004 that checks for text similarity between a text of interest – say, the one I want to submit for publication – and the universe of published works. At the time of writing this the second book edition, iThenticate was the predominant plagiarism detection service used by scientific publishers and funders. Universities, such as my own, require graduate students and mentors to run dissertations through iThenticate as a means to assure these documents are plagiarism-free. Since iThenticate is an important feature in the world of scientific publishing, I thought it would be interesting to "put it through its paces" as it were and to show how it works. To that end, I wrote a brief grant proposal that includes purposefully plagiarized text. In the following pages, I show the proposal as it would be read by the funder, an "illuminated" version where I show with bold font where I know I plagiarized, and then several screen shots of iThenticate output.

Case study: grant proposal that includes plagiarism

In this case study we'll see how iThenticate finds text similarity, i.e., plagiarism.

A CRISPR/Cas9 gene drive system with induced sterility engineered into weedy plants to defeat herbicide resistance: a brief sci-fi narrative that should definitely not be implemented, which also purposefully contains plagiarism.

C. Neal Stewart, Jr.
Department of Plant Sciences
University of Tennessee
nealstewart@utk.edu

Weedy and invasive plants are among the most damaging pests in cropping systems. It is no secret that weeds are farmers' worst pest, causing up to US$20 billion in losses every year (Bridges 1994). We now potentially have the tools to combat weeds using their own genetics and tendency to spread "like weeds." A potential solution is to engineer a gene drive into weedy plants using a genome editing tool such as CRISPR/Cas9 that would turn the weediness of weeds upon themselves. Clustered regularly interspaced short palindromic repeats (CRISPR) is emerging as a plant genomics and biotechnology research juggernaut. Gene drives are most prominently being developed by companies such as Oxitec to control mosquitoes. Oxitec uses advanced genetics to insert a self-limiting gene into its mosquitoes. The gene is passed on to the insect's offspring, so when male Oxitec engineered mosquitoes (males that don't bit) are released into the wild and mate with wild females, their offspring inherit the self-limiting trait. The resulting offspring will die before reaching adulthood, and the local mosquito population will decline. Similar technology could conceivable be used in weed biology. In the Oxitec system, are sterility genes that become operational to then torpedo the local mosquito population, which would serve to decrease the primary vector spreading malaria.

Weed genomics, while still a nascent field, has given us some molecular tools to understand the genetics of weediness (Basu et al. 2004). In addition, there have been some genetic engineering methods developed to introduce a gene drive into weedy species such as *Conyza canadenisis* (horseweed) (Halfhill et al. 2007) a widely distributed agronomic "malophyte." So, we have most of the tools needed to implement a gene drive in weeds. How it might work follows. A gene drive could be transformed into a glyphosate-resistant biotype of horseweed, which would carry a fitness advantage, enabling the resistance to spread throughout a population. Carried with CRISPR/Cas9 glyphosate-resistance allele will be a second gene with induced by a promoter induced by a chemical agent and that directs expression to pollen, which results in pollen ablation by the expression a restriction endonuclease gene (Millwood et al. 2016). Thus, after the population increases after one or two growing season, all seed production could be eliminated by inducing male sterility in the population, hence no seed production to compete with next year's crop. The gene drive would then effectively combat the growing problem of the evolution of herbicide resistance in weeds and be an effective management tool. If gene drives successfully eradicated these invasive populations, many would rejoice. It is anticipated that no spreading of the gene drive or effect would occur beyond the target population, even though invasiveness of the system would be initially desirable. Although resistance prevents drive systems from spreading to fixation in large populations, even the least effective systems reported to date are highly invasive. Releasing a small number of organisms often causes invasion of the local population, followed by invasion of additional populations connected by very low gene flow rates. Examining the effects of mitigating factors including standing variation, inbreeding, and family size is necessary for containment. The proposed research could solve the problem of weeds evolving resistance to herbicides such as glyphosate.

References cited

Basu, C., Halfhill, M. D., Mueller, T. C. and Stewart, C. N. Jr. 2004. Weed genomics: new tools to understand weed biology. *Trends in Plant Science* 9: 391–398.

Bridges, D.C. 1994. Impact of weeds on human endeavors. *Weed Technology* 8: 392–395.

Halfhill, M. D., Good, L. L., Basu, C., Burris, J., Main, C. L., Mueller, T. C. and Stewart, C. N. Jr. 2007. Transformation and segregation of GFP fluorescence and glyphosate resistance in horseweed (*Conyza canadensis*) hybrids. *Plant Cell Reports* 26: 303–311.

Millwood, R. J., Moon, H. S., Poovaiah, C. R. Muthukumar, B., Rice, J. H., Abercrombie, J. M., Abercrombie, L. L., Green, W. D., Stewart, C. N. Jr. 2016. Engineered selective plant male sterility through pollen-specific expression of the *Eco*RI restriction endonuclease. *Plant Biotechnology Journal* 14: 1281–1290.

What iThenticate found as plagiarism

During the creation of this fictional proposal, I intentionally plagiarized to see if and how iThenticate would show the text similarity between my proposal and published work. As we can see in Figure 3.2, there are four instances of text similarity on the first page. The first is my name and affiliation, which can be ignored, but the other three are instances of plagiarism. We see that iThenticate lists (on the rightmost bar) the plagiarized source. If we clicked on the source, it would show the reader how the similar text maps back to the original source. In these instances, the information in my text is specialized information, and therefore an appropriate reference should have been supplied in each case. Figure 3.3 shows some breakdown of the plagiarism. In the first instance ("it is no secret …"), I self-plagiarized a review paper I coauthored (ss et al. 2004) but didn't cite it in this instance. Instead, I cited a reference internal to the Basu paper. That practice of simply copying and pasting with an internal reference is not uncommon, but not allowed. The second instance ("is emerging

Figure 3.2   In this purposefully plagiarized grant proposal document, iThenticate was used to find text similarity and track the source text to publication. The author bolded text that he copied and pasted into the grant proposal.

as a plant genomics juggernaut") is also self-plagiarized, but is not as serious in that it is what anything in plant biotechnology would know, but still should have been cited or put in quotes. The third instance of plagiarism is the most interesting on this page since I copied from a company's webpage. Note that I misspelled "bite" as "bit," which would potentially tip off the reader that there was mixed writing. When we go to the next page of the analysis (Figure 3.4), we

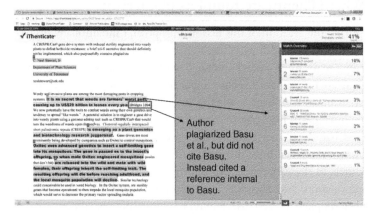

Figure 3.3   The example shows the case where text was copied and pasted but did not cite the reference where text was copied. Instead, a reference internal to the source text was cited, a clear indication that plagiarism was intended.

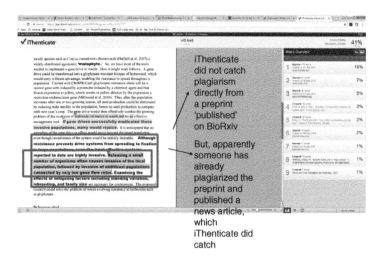

Figure 3.4   The example shows a case where only part of the plagiarized text was detected by iThenticate. It also shows a case of apparent secondary plagiarism.

see that iThenticate did not "find" all the bolded text that I copied from a preprint. Preprints are simply manuscripts that may or may not have been submitted for publication in a real journal. Preprints such as the one I used to copy text are curated in preprint servers such as BioRxiv. I was surprised that the preprint text was not "crawled" by iThenticate. Another surprise was that preprint text had already apparently been copied and used in a news article, which iThenticate did find. Thus, this simple case study shows that plagiarism detection software, while very good, is not perfect at finding text similarities.

Self-plagiarism and ethics revisited

The accepted definition and repudiation of plagiarism is clearer than that of self-plagiarism, although there is strong agreement that self-plagiarism should be avoided. Judicious recycling for unpublished sources is allowable. For example, if I am asked to make four presentations in one year to different audiences about the same topic or study, I am probably going to recycle parts of the abstract and subsequently update and refine it. To make a totally new abstract each time is a waste of energy and not useful to science or society. There is agreement that self-plagiarizing papers and proceedings is very detrimental to science (Green 2005; Berlin 2009). However, allowances should be made for refinements of published proceedings articles to morph into peer-reviewed journal articles. Self-plagiarism should be avoided and science and society benefit. Ethical considerations of harm and benefit should be minded as we perform writing assignments.

Judge yourself

✓ How do you feel about your submitted manuscripts being scanned computationally for plagiarism?
✓ How would you feel if plagiarism was detected and your university administrators were alerted?
✓ Do you have a trusted colleague you can ask to read drafts of your papers who will make an "off-the-record" judgment about your writing and instruct you on plagiarism?

Is plagiarism getting worse?

I present no data here, but I think the consensus answer among researchers is generally "yes," with plagiarism being greatly enabled by full-length online articles and the internet. Plagiarism is to academics as steroids are to sports. Cheating becomes easier because of technology, and there seems to be an escalation in prevalence and perhaps tacit acceptance. For example, many journal editors look the other way when it comes to copied methods sections. Also, I've heard plagiarism defended as acceptable because of the difficulty of non-English speakers having to communicate their science results in English, but this is not an acceptable excuse. As in most technology-related crimes, detection is always a playing catch up, but it is heartening that more and more journals are using tools like iThenticate so they can avoid publishing plagiarized material.

Case study: Tricky Dicky the plagiarist

In the 1980s, it had been reported that a graduate student (Dicky) in political science had allegedly plagiarized on a term paper for a course. The number of enrolled students in the course was large and Dicky sat toward the front of the class on the rare occasion he attended. He sat in front because he was perpetually late and those were the only seats available by the time he arrived. Dicky was a snappy dresser and drove an exotic sports car. He was the rare PhD student with much money. While Dicky imagined himself as a scholar, he enjoyed playing golf more than pursuing his studies.

This case occurred just before the introduction of the World Wide Web, when word processing was done on mainframe computer terminals and the results were printed off remotely in the computer science building. Therefore, typewriters (and typists) were still alive and well, and the industrious student who was endowed with fast fingers could earn some extra cash as a typist. Typically, the trade was advertised on bulletin boards with rates and phone numbers listed on tear-offs at the bottom of the page.

Dicky happened to, unbeknownst to him, choose such a typist to employ who just happened to sit in the middle of Dicky's political science class: Sally. In addition to being a fast and accurate typist, Sally was also a very good political science student and an observer of people. Dicky, being neither observant nor often present in class, did not know or recognize Sally. Therefore, when Dicky met Sally for the first time, it was to hand her a composite of typed pages that were cut into sections and taped together, with the instruction to type this "for a friend." While Sally might have inquired about the unusual nature of the request and its propriety, being the consummate professional, she did not. She merely received the pages and did her service. She did, however, recognize her new client and the job for hire as the assignment that had been given three weeks prior in their class.

In a week, when the assignment was completed, Sally made a photocopy of her typed work-for-hire, and then gave Dicky the original, which he promptly turned in as his term paper. As one might imagine, by the week's end, Sally had grown angrier and angrier at Dicky; the result of having typed her own term paper that she wrote herself as well as Dicky's, which was, of course, plagiarized. Dicky was clueless, of course, until he was notified that he was being investigated for plagiarism. Dicky was incensed that Sally had spilled the beans, and he offered no real defense. He received an "F" for the course and probation. Within the year, however, he had withdrawn from the university and switched fields.

This story is illustrative inasmuch that Sally would not be contacted today to type Dicky's project, and therefore there would be no one to report Dicky's bad behavior. The reader would have had to pick-up on disparate writing styles to detect. The Dickys of the world know this and plagiarize all the more. I'm certain of it. All they have to do is cut and paste, just as Dicky did, but using a computer, and there is nobody who knows they are doing it.

The [true] case study: the plagiarizing novelist who also plagiarized her confession to plagiarism and the author of the website "Plagiarism Today"

Perhaps the most bizarre case of plagiarism I've been made aware is that of Jumi Bello, documented by Jonathan Bailey, the author of the Plagiarism Today website (http://plagiarismtoday.com). It seems that in Ms. Bello's her first try at writing a novel, "The Leaving," she plagiarized various lines of text from other writers. An unfortunate consequence for this budding novelist was the cancelation by the publisher when they discovered the plagiarism in 2022, just months before *The Leaving* was to have been published. Apparently, Ms. Bello then felt the need to explain why and how she plagiarized. She did so on the Literary Hub (http://lithub.com), which almost immediately retracted the plagiarism atonement piece because of inconsistencies and [ahem] plagiarism. In her essay, she divulged information about her own bout with mental illness, the pressure of hitting her book submission deadline, and how she dealt with this pressure. Moreover, she describes how she writes: essentially using the copy and paste method, in which she subsequently employs paraphrasing. With her book manuscript submission deadline looming, she forgot to paraphrase. This is how she said the plagiarism happened. In the explanation essay, she harkened back to the first putatively documented case of plagiarism in which the Roman poet Marcus Valerius Martial had been plagiarized by Martial wannabees in the first century AD. While it was an interesting approach to go back to the beginning of appropriating others' material without citing them, she plagiarized the Plagiarism Today guy who'd written about poet Martial a decade earlier. The cases are documented here: https://www.plagiarismtoday.com/2022/05/09/plagiarism-today-plagiarized-in-a-plagiarism-atonement-essay. Or in the words of songwriters Max Martin and Rami Yacoub, "Ooops! ... I did it again." It appears that in her "why I plagiarised essay" Jumi Bello had copied parts of Jonathan Bailey's earlier writing and insufficiently paraphrased Bailey and not given him credit. I've never seen a more illustrative example of how sideways the copy and paste method of writing can go and the reasons why to not practice it. Or in the words of Bailey: "The way you avoid plagiarism isn't to 'change the language' but to never have that language in your original work in the first place" (https://www.plagiarismtoday.com/2022/05/09/plagiarism-today-plagiarized-in-a-plagiarism-atonement-essay).

Summary

✓ All plagiarism and self-plagiarism is unethical and should not be practiced.

✓ Some text recycling from your previously unpublished writing is acceptable.

✓ Computational tools are being increasingly used and will catch serial plagiarizers.

Chapter 4

Finding the Perfect Mentor

ABOUT THIS CHAPTER

- The choice of a mentor, especially as major professor or postdoc supervisor, is a crucial decision for young scientists.
- Deciding on a mentor is more important than the choice of institution.
- There are several signs of good and bad mentoring that can inform the decisions of students and postdocs.

The perfect mentor doesn't exist, of course, but a good mentor can help launch a young scientist's career. A bad one can kill or cripple it and encourage mediocrity. This chapter is written mainly for the scientists at the beginning of their careers: to help in the selection and then, "training," of a mentor, and to provide guidance and tips when someone subsequently becomes a mentor, which is bound to happen when scientists run their own labs. But actually, the point in time when that someone actually becomes a mentor is open to interpretation. I'd like to think that graduate students and even undergraduates can aid the mentoring process in helping more junior members of the lab gain useful experience and savvy. I've observed this in action with great results. The ethics of mentorship are complex. On one hand, laboratory directors (mentors) must be selfish in one sense in that opportunities must be prioritized to maximize results. Building one's own career and reputation must also be a consideration, no matter the stage or age of a scientist. On the other hand, lab directors

---

Research Ethics for Scientists: A Companion for Students, Second Edition. C. Neal Stewart, Jr.
© 2023 John Wiley & Sons Ltd. Published 2023 by John Wiley & Sons Ltd.
Companion website: www.wiley.com/go/stewart/researchethics2

have an ethical obligation to nurture graduate students, lab staff, and postdocs: the people who actually do the experiments, write the papers, and even participate in grant proposal writing. Good mentors find no dichotomy in performing the best research where trainees are nurtured and mentored so that someday they too, can fledge into their own labs.

My view of mentorship is likened to one-on-one teaching and enabling junior mentees to also mentor. Mentorship also includes the relaying of professional ethics to the less-experienced scientist and especially how to become an independent researcher, which includes a plethora of "secret" rules and nuances. There are many ways a mentor can go wrong, and the student has to be proactive in choosing and keeping a mentor active in good mentorship. Postdocs and graduate students must engage the mentor to communicate what they need. The mentor is the hinge that swings the door for young scientists-in-training. But the mentee often has to push the door open. As doors open or close, it becomes clear that choosing a mentor is one of the most important decisions to be made in science, even though often it is greatly underappreciated and overlooked.

Caveat

Ok, I don't think I'm super mentor, with empathy faster than a speeding bullet and insight able to leap tall personnel problems with a single bound. But, good mentors abound, and I am certainly grateful for mine: those scientists who supervised me during graduate and postdoc training, but also those people who helped me learn the systems inherent to science and universities. There is no shortage of peculiar politics in these institutions. Many successful scientists tell stories of having successful mentors. Much of the information in this chapter arises from my own mentoring mistakes and not those from my own mentors. My material also comes from observations of mentoring and stories from graduate students and postdocs – not many of my own – but those of others. Names and places are changed. Their stories are in *italics*.

The first step a graduate student or postdoc faces is choosing the right mentor. It also is increasingly common for assistant professors to be assigned a faculty mentor to help them become accustomed with their institution and department, as well as giving them advice on tenure and promotion, so this chapter is applicable to a wide range of scientists in various stages of their careers.

Choosing a mentor

Aside from the chosen field of science and a potential project, the mentor is probably the most important factor that shapes the initial direction of a scientist's career. The trick is to find the mentor working in a setting and doing research that is compatible with and compelling to the student. And hopefully the mentor also has substantial funding! Science without funding is like a car without petrol; not moving. Graduate students in science are nearly always paid a stipend to help them obtain their advanced degrees, but money is not always universally available. Finding the best major professor suited for a particular candidate has been greatly aided by the internet, and fortunately, it has never been easier to locate and correspond with candidate professors. On the other hand, faculty members are inundated with emails, sometimes not personally addressed, requesting opportunities for study. A universal process for finding graduate school opportunities does not exist, but generalized guidelines for finding an advisor are straightforward.

Home away from home: the university, department, and program

Making the optimal choice for all three of these items above important – just not nearly as important as choosing a mentor. A student will be linked forever to the university where the PhD was earned. Whenever you give a guest lecture, seminar, or any other presentation, you will hear the name of the university as part of your introduction. Your PhD hood that you'll wear for university commencements is identified with the university and college. Sometimes the PhD robe is also unique (e.g., the crimson robe of Harvard University is always easy to spot in graduation exercises). Your reputation and the reputation of the university and department/program are indelibly linked, making these important considerations. But I argue here that the importance of finding a good mentor outweighs the reputation of programs or universities, because the mentor, unlike an institution's reputation or facilities, directly shapes the student in myriad ways.

Some students want to attend a particular school for location, reputation, amenities, and other reasons (e.g., "my dream is to have a degree from 'Exclusive U'"). While these justifications are not invalid, they add unnecessary constraints that can muddy the waters of deciding which lab to join. Not to fear: the best mentor will seldom be located at the worst university or worst department. I think the best search scheme to find major professor candidates is to determine who are the top 5–10 scientists in your particular research area of interest, and then match the entrance requirements of their associated department and program with your own qualifications. After all, it is the major professor's lab where the student spends the most time, and it is the major professor who will be giving out advice and shaping the research project and subsequent publications. The quality of the science or engineering performed coupled with mentoring will be the major determinants of the next step in a career, along with the success enjoyed. In many programs, the major professor also has a strong voice on which students get admitted via funding and advocacy.

The major professor

Start early. Choose wisely. Be persistent. In the best case, the choice of major professor can propel a career to the stratosphere. The student will be first author on multiple major research papers published in excellent journals, and will learn, first-hand, what a successful career in science looks like. In the worst case, the wrong choice can drive a student out of science, or worse. One achievement that pleases me most is that the overwhelming majority of my ~200+ trainees is still in science and enjoying fulfilling careers.

The first step toward being a scientist is knowing the field of interest by reading pertinent journal articles. Computer-based literature searches of keywords to find journal articles that are interesting are a good first start. Once you find articles of interest, determine which person in the author list is the professor driving the research. Usually papers have several authors, but few of these are professors. Typically, but not always, the principal investigator (the "PI") is the "corresponding author" (in biology, often the last author) of the paper.; The PI often has participated in the ideas behind papers. It is important to study with people with good ideas. After a list of prospective major professors is made, the next step is to find potential mentors on their university websites, typically listed by departments. Then read web pages and represented papers authored by the person. Short of stalking the major professor, you must learn all you can about this person. Students should respect and be interested in the mentor's research, and then it helps to actually like the person too.

A big mistake is to enroll in graduate school or a take a postdoc position without actually visiting the prospective mentor and the facilities. There is no substitute for setting foot in labs and offices looking eyeball to eyeball with potential mentors. The COVID-19 pandemic and other troubles sometime make an in-person visit difficult, but if at possible, visits should be made. During a one-on-one meeting with the professor, it is important to gage how this person will perform as *your* mentor. For example, how does this professor deal with interruptions? Is the prospective mentor warm or cold? It is imperative to learn how students are funded, and for how long. Beware of the PI using the phrase "my grant" or "my grant proposal" as in the singular. A lab that operates on a single grant is typically not stable funding-wise since the grant might or might not be renewed every three years – the research might not continue. Where does this leave the student? Of course, many faculty members in the United States have National Institute of Health (NIH) R01 grants that seem to be renewed forever; that funding is typically quite stable. Certainly, the PI's philosophy and practice of authorship should be discussed. And most importantly, the student should meet and talk with other graduate students in the lab in an attempt to determine how pleasant and productive your experience will be if you join. Facilities and equipment are important as are materials and supplies. If budgets always tight, then students and postdocs will be frustrated. During an interview, you really want to determine if you can envision yourself working in that setting with the people you meet. Lab colleagues are nearly as important as the major professor, and in certain cases, more important. You'll work with many of them shoulder-to-shoulder each day while the boss is sequestered away in an office or on travel.

"When I arrived to his office, he was fiddling around with his computer and was obviously frustrated with some software glitch. He begged for me to be patient while he fixed this one problem prior to our scheduled meeting. His frustration level and temper rose until he finally abandoned his computer issues and turned to me. His tone suddenly changed as he went from frowning to an eerie smile.

The smile reminded me of a character from a vampire movie. With that we began to talk about research, which I found quite interesting: both in theory and methodology. We were having a good conversation and I was imagining myself working with him. Then, his graduate student showed up at his door and simply stood there. She seemed to be expected but she did not interrupt our conversation. Finally, he noticed her and, his demeanor changed again. He asked quite tersely for her to pick up four specific books from the library that he needed for a grant proposal. 'Make sure they are on my desk before lunchtime' he said. Then he returned his gaze upon me once more accompanied by that eerie smile. After the books arrived and our meeting concluded, he took me out to lunch and promised me a good stipend and the project of my choice. It should have hit me square in the face, but it was only on my way home that I realized that something was not right with his lab, and that it would be a mistake for me to join his group. I guess I would not want to be treated the way he treated his current student. And I wouldn't want to spend the next few years with a Jekyll and Hyde type of boss, or worse, a vampire who'd want to suck me dry. Getting a PhD can be something of a horror. Who needs the unnecessary drama?"

It pays to heed the vibe! It is smart also to determine whether a PI's current students and postdocs are happy in their research or whether, given the chance, if they would choose another mentor. Other questions are worth asking. Do students graduate with their intended degrees? And how long does it take? Are people happy? Where do people go after they graduate? Where do postdocs typically go once they leave the lab? Are they still in science. Life in science is – surprisingly to many people on the outside – predominantly about people and their interactions, namely the mentor–student and lab member–lab member relationships, which shape the environment and success of the mentors' labs and their trainees. Ethics and best practices can be affected by the lab vibe. "The potential for the breakdown of trust underscore the need to look more closely at relationships in research and laboratory groups, and at the ways in which groups are managed. A research environment in which relationships are distant, frayed or fractured – 'an unhappy lab' – may well not sustain responsible research conduct. Trust is of key importance to the enterprise of science" (Weil and Arzbaecher 1997). Potential trainees have to realize that the professor is the person who most controls the lab vibe and successes of the people who are members of the group.

Judge yourself

✓ How good of a judge of character and environments are you? Can you "size-up" people? How rapidly?

✓ Are you an "easy sell" or able to resist promises that are too good to be true?

✓ What are your most important desired features in a mentor and laboratory? How does your personality and style play into your choices?

✓ How do you feel about mentor accessibility? Do you prefer to walk in and chat when someone's door is open or would you rather make an appointment? Are you a texter or an emailer? What about your mentor?

✓ Are you a trusting person? Do people tend to find you trustworthy in return?

The two big things in deciding on a mentor: funding and papers

The most important scientific metrics are associated with productivity, namely the BIG TWO: grants and papers. If a mentor can encourage a trainee to become independent in these two items, and to approach them with respect and integrity, then good mentorship is occurring and there is a good change the trainee will have a fulfilling training experience and career.

Authorship

It is crucial that there is a clear understanding about a mentor's style and practice about funding and authorship. In a perfect world, graduate students are first authors on any papers stemming from their graduate projects and be middle authors on side projects in which the students have made significant contributions. Alternatively, some major professors prefer to take complete control of papers and thereby deny students opportunity for independence and growth. There are quite of few PIs somewhere in the middle of this continuum. Related to this, prospective students and postdocs should determine if professors prefer to publish lots of shorter papers with few authors or fewer grand papers with many authors, or again, something in between. One scenario is that several graduate student projects might be needed to make a grand paper that might be mainly controlled by a postdoc or the PI who would then likely be the first author. Never lose sight that the students and postdocs should protect their interests in publications as authors. At the same time, there may be compromises needed for the greater good, and co-first authorship is becoming increasingly common. The bottom line is that for young scientists to develop in science and progress to the next career stage successfully, that the number and quality of papers authored and authorship position are very important factors. Papers should be a reflection of the student's expertise, research productivity, and real contributions.

Funding

For many graduate students and postdocs, research funding "just appears" and is never a worry. In some unfortunate cases, funding is not stable wherein a student is not aware of the magnitude of instability until the money is gone and a student is "self-funded," i.e., not funded at all. I have witnessed PhD students abruptly dropped when the PI loses his funding. It is a terrible conundrum for the student. As a mentor I need to make sure funding is stable. That said, one happy medium is between mentor as Santa Claus and lab poverty is the one in which graduate students and postdocs play active roles in obtaining funding and are partly responsible to report research progress to agencies or sponsors. If young scientists are to be successful in research, then they must become familiar with the world of grants and contracts. What better place to start tuning-in than as a PhD student, and then continue to fine-tune grantsmanship during a postdoc stint? Many mentors require trainees to produce portions of proposals that get rolled into a larger document; mini-proposals of sorts. In other instances, young scientists can land their own small (or not-so-small) grants and become masters of their own universe. Landing my first (and second and third!) small grants while I was a graduate student were as joyous to me then as the big grants I get now. At the time, $600 from Sigma Xi and $1000 from a company was crucial to paying for supplies and a computer for my PhD project. At the time, I was on a nine-month teaching assistantship and was working on a project that was otherwise not funded with a federal grant. My early activity in fundraising also gave me confidence in playing the grant game. I learned, I liked writing grant proposals, and with practice I became pretty good at it. In fact, success in grant-getting is quite often tied to winning tenure in many colleges and universities where research is valued.

It is important to keep in mind that, ideally, the project you adopt as your own should be fundable, if not already funded. Science is fueled mostly by government grants. In the United States, the NSF and NIH fund a miniscule portion of grant proposals submitted each year. Therefore, successful scientists must necessarily be proficient at getting their grant proposals funded. It is fair game to ascertain how proficient your potential major professor is in the grant game. Ask for a list of awards. Shy away from someone who won't give you a list of grants or talk about funding. I've seen grant ruses played many times and the student is the one who gets pinched. I know of a faculty member who has claimed he's had multiple grants from a certain agency and a search of the agency's database fails to turn up any grants where he was the lead PI. I've seen faculty members abruptly lose their funding and students are immediately dropped with no warning (and no apologies). Students must practice due diligence to parse the fakers and dilettantes from the real deal. Doing fundable research is your ticket to having a chance at being a successful and independent research scientist. Students burdened with unfundable research typically will not be afforded the opportunity to reach their potentials. Some of the best advice I ever received as a graduate student pertained to the nature of my research and how I thought about research at the time. A professor on my graduate committee frankly informed me that my masters research, while probably publishable and interesting to me, was definitely not fundable nor of interest to a wide array of researchers in my area. He also offered that people would pay me to be interested in several other topics, so I stretched my own interests to pursue something more fundable for my PhD, and then even more fundable (funded even!) for my postdoc research. A teaching assistantship allowed me to follow my nose in seeking research topics and my mentor provided a lot of latitude in research. This stretching put me out of my comfort zone but also led to my gaining new skills and expertise. Research is typically a winding path, and not a highway; this is ok. Successful scientists are able to master new techniques and develop their own thinking. The successful graduate student is adaptable and grows up to be an adaptable senior scientist. When beginning a graduate degree, it also helps if adaptation happens rather quickly.

> *"One of my best friends in graduate school was so much smarter than me. In biochem, she was the first one to finish homework sets (and hers were always right). Her grades were stellar. I felt really bad for her when they kicked her out of the graduate program after her second semester. They gave her the boot because she would never decide on a PhD project. Picking a project, selecting a committee, writing a research proposal, and giving a talk about the proposal were all things that were required to be completed before the end of the second semester. The big stumbling block was deciding on a single project. She was just indecisive I guess … and then she was gone."*

Final thoughts for graduate students and postdocs in mentor selection

There are a few things and additional items to consider when making a selection on a mentor. It is a measured risk to select a beginning assistant professor as a mentor. The risk is that the mentor does not get tenure and is fired from the university. Typically, the system demands that the assistant professor is either "up or out"; i.e., getting tenure and a promotion to associate professor or out of a job. In the case of the latter, the graduate student typically becomes a sort of ward of the university and must scramble to finish a degree, and the postdoc is simply out of a job. So, it is wise that before trainees sign on with an untenured faculty member that they are convinced that their mentor will be awarded tenure. If there is any hesitancy from the assistant professor about their chances in getting tenure and promotion to associate professor, then buyer beware! If an associate professor or full professor simply changes jobs, there is usually good opportunity to remain with the mentor at a new location. When I moved to my present position, nearly my entire lab elected to move with me and many of these people in turn attained significantly better pay and project support. None fared worse. On the other end of the spectrum, you'll also want to keep an eye out for age, health, and retirement prospects of a potential major professor. Your career will not be accelerated if your mentor retires, dies, or spends lots of time in the hospital during your graduate program.

There is also the consideration of mentor success and lab size. Malmgren et al. (2010) performed an analysis on mathematics mentors and the fate of their trainees. They found that trainees of mentors with lots of trainees had more trainees themselves. They also observed a correlation between number of trainees and productivity. Another interesting result was that there was a benefit to being a trainee of the young successful professor compared with the older successful professor. Therefore, it is beneficial for a student to pick a successful mentor, but there is a greater benefit to picking the mentor early in the mentor's career. I'm not sure how a person can know very early in a career how much a mentor will succeed later, but this study confirms what many people have suspected all along in mentorship: success breeds success.

Harassment and bullying by mentors

A final, and, definitely, unpleasant topic in this section is avoiding potential mentors who have negative reputations and behaviors. Many workplace relationships are asymmetrical, in which one person has substantially more power than another. Since science training is largely an internship of sorts, and mentors are regularly tenured with a job-for-life guaranteed, there is built-in potential for abuse. In recent years, harassment and bullying by mentors have become uncovered as a potentially devastating and widespread problem for trainees. Indeed, in a survey performed by Nature in 2020, which included 7600 responses from postdocs from 93 countries, bullying and power imbalances was the most often-reported item experienced (65% of respondents) or observed (74% of respondents) (Woolston 2020). Below I will examine some prominent publicized instances of sexual misconduct/harassment and bullying and the responses of institutions.

The National Academy of Sciences of the United States is the most prestigious collection of membership-elected scientists in the United States. Election to the National Academy is for life. In 2019, the National Academy revised their rules to allow for members to be expelled for sexual harassment and other kinds of misconduct. In 2021 two members were expelled after very public cases were evaluated at their respective universities. Geneticist Francisco Ayala was accused of harassment by four women – from an assistant dean, to faculty members, to a graduate student – at the University of California-Irvine. A committee there found that he had indeed broken policy at the university by words and deeds that made women feel degraded. These actions included inviting a faculty member to sit on his lap at a research seminar, his inappropriate touching, and his making sexually explicit comments. He was terminated from the university because of these findings. In another case, astronomer Geoffrey Marcy who had been employed by the University of California-Berkeley was the second National Academy member expelled in 2021. He'd previously resigned from his professorship at Berkeley after he was found to have breached university policy on harassment. For both Ayala and Marcy, there had been decades-long histories of complaints and warnings to them to stop offensive behaviors. In both of these cases, US law Title IX – which prohibits sex discrimination at institutions that receive US funding – played an important role for the two California universities to investigate and apply findings that essentially led to the ouster of the famous professors from the National Academy.

Bullying has gotten a lot of attention in recent years since it may be considered "endemic in academia" (Gewin 2021). Unlike sexual harassment, which has been carefully defined throughout the years, bullying is a more amorphous set of behaviors. As noted by Gewin, Morteza Mahmoudi at Michigan State University cofounded an organization to expose bullying and defines it at "hostile behaviours that include ridiculing, blaming, invasion of privacy, and put-downs." Probably everyone reading this book has been bullied and has been a bully sometime in life, so we all know it when we see it and we all know how bad it can be and feel. It is really the sustained bullying that is truly problematic and especially so when perpetrated on people that the bully has direct power over. The bigger the power differential, the bigger the potential problem. Gewin says that the first step in confronting an academic bully is to confirm that the activity in question is indeed bullying. She provided examples such as blowing up about relatively small mistakes and threatening to remove a student's name from a paper. The second step is seeking support, and the third is formal and informal complaint routes. These routes differ widely among institutions and countries. Depending on the situation, an informal resolution process may be quicker and more effective than pursuing official complaints. Therefore, it is a safe assumption at the time of writing the second edition of this book, that egregious cases may be the best candidates for formal resolution, which may also result in sanctions against the bully. One such high-profile example is the Max Planck Institute for the Science of Human History director who lost her job over bullying: Nicole Boivin; twice, in fact (Curry 2021). The bullying in question purportedly included taking credit for others' work. After she was demoted in 2021, a court overturned the decision, which triggered Max Planck to demote her a second time in 2022. According to Abbott (2022), a total of six other Max Planck directors have either been demoted or investigated. These cases have triggered various critics to question if the investigations were biased and whether the Max Planck Institutes' decisions have been fair (Abbott 2022). Indeed, it appears that there are few "clean-cut" cases of university-punished pure bullying in the literature. Most of the bullies that are subsequently fired were also found to be guilty of research misconduct and/or sexual harassment. Probably the most prominent recent case was that against Eric Lander who held the position of Science Advisor to the President (Biden) for around a year. He was ousted from that position for bullying people who reported to him. Prior to that job he was a professor at Harvard and MIT for decades. If he had the reputation for abusive and bullying behaviors prior to the presidential appointment, it is unclear if there was ever any ramification stemming from these activities while he was a faculty member. Prior to his presidential appointment "500 Women Scientists" (2021) objected to his nomination on the basis of certain controversies that they believed downplayed the contribution of women in science, which had multiple points view depending on perspective. I suppose my point here is that bullying is in the eye of the beholder, somewhat ill-defined, and probably part and parcel of academia to some degree. My guess is that any large organization would have a difficult time totally removing bullies and bullying from their ranks. I do see one solution to get the worst of bullies fired. In 2018, Nazneen Rahman was fired for bullying activities that spanned over a decade at the Institute of Cancer Research (Weale 2018). There are a lot of bullies in academia, but few bullies are accused of bullying by 45 current and former trainees in a letter to their institution. It appears that theirs was not the first set of allegations and that many had come before with few results. So, my advice to potential victims of bullying is to try to read the situation prior to joining the erstwhile bully's lab. And if you're in such a toxic place as Dr. Rahman's lab, commiserate and write a letter that the administration of the institute cannot ignore.

Choosing a graduate project

Certainly, the consideration of what project to do for masters and PhD research is as important as choosing a mentor. Sometimes in choosing a mentor, the choice of project is already made. That is, the mentor might not have any choices for a new student than the one specific project prescribed by a grant. In most cases there is some flexibility. As Reis (1999) quotes department head George Springer, "It is really important to do the right research as well as to do the research right. You need to do 'wow' research, research that is compelling, not just interesting." I think this is really good advice in that it inspires students to aim high and not sell themselves short of their potential. Of course, this book is more about doing the research right than doing the right research. What project is the right project depends on many factors, especially the individual student. At the end of the chapter, there is a case study that works through many of the relevant factors.

Judge yourself

✓ How aggressive or assertive are you? Do you feel like you take control of situations without being a bully?
✓ Are you a good judge of character?
✓ Do you have trusted friends and colleagues whose opinions you trust?
✓ Do you listen to advice?

Mentors for assistant professors

In many departments, new assistant professors are assigned faculty mentors who are established and experienced. In many ways, mentorship to assistant professors is far less crucial, career-wise, than for postdocs and graduate students. Why? Since obtaining an assistant professor position is probably the biggest employment bottleneck in all of science, anyone who has gotten that far has probably already received valuable mentoring. Therefore, being found worthy of filling an assistant professorship indicates a person has enjoyed a certain amount of success and independence already that is sure to spill over into working as a faculty member and earning tenure, which is generally related to having success in grants and publications. A sage mentor or group of mentors can help an assistant professor land that first big grant and become even more productive and independent. Alternatively, a bad mentor or a toxic department can inadvertently drive the mentee to another university or out of science. I have seen both the good and bad scenarios in action. It is paradoxical and unfortunate that some departments acquire the reputation of "eating their young" by bad mentoring. No one wants to be eaten. This figuratively occurs when mentor(s) attempt to make assistant professors their minions: underlings who can write their grant proposals, papers, and other things the senior faculty have grown too lazy to do or things they are no longer competent to do since science has long passed them by. *At any level, good mentors always make their trainees increasingly more independent and not the other way around.* This, in fact, is the key outcome of good mentoring. I believe that it is only when scientists or engineers are truly independent can they become valuable colleagues and collaborators. Again, assistant professors should choose their mentors wisely. Ask the department chair about which faculty members in your subfield are the very best mentors. Talk with faculty members who were mentored by the candidate offered by your chair. Better still, ask if you can be assigned a team of mentors. With a team, you can better manage the one you might discover who seems hungry for a minion.

Laws of Herman

Irving P. Herman is a physics professor at Columbia University, who offered advice to graduate students through an article that appeared in *Nature* (Herman 2007).

While he states that these "laws" are "slightly exaggerated" I think that they are the kind of advice a good mentor gives. In addition, he offers his advice in a humorous and disarming fashion.

1. Your vacation begins after you defend your thesis.
2. In research, what matters is what is right, not who is right.
3. In research and other matters, your adviser is always right, most of the time.
4. Act as if your adviser is always right, almost all the time.
5. If you think you are right and you are able to convince your adviser, your adviser will be very happy.

6. Your productivity varies as (effective productive time spent per day)^{1000}.
7. Your productivity also varies as 1/(your delay in analysing acquired data)^{1000}.
8. Take data today as if you know that your equipment will break tomorrow.
9. If you would be unhappy to lose your data, make a permanent backup copy of them within five minutes of acquiring them.
10. Your adviser expects your productivity to be low initially and then to be above threshold after a year or so.
11. You must become a bigger expert in your thesis area than your adviser.
12. When you cooperate, your adviser's blood pressure will do down a bit.
13. When you don't cooperate, your adviser's blood pressure either goes up a bit or it goes down to zero.
14. Usually, only when you can publish your results are they good enough to be part of your thesis.
15. The higher the quality, first, and quantity, second, of your publishable work, the better your thesis.
16. Remember, it's your thesis. You (!) need to do it.
17. Your adviser wants you to become famous, so that he/she can finally become famous.
18. Your adviser wants to write the best letter of recommendation for you that is possible.
19. Whatever is best for you is best for your adviser.
20. Whatever is best for your adviser is best for you.

The Laws of Jason: mentorship from a postdoc's perspective (contributed by former graduate student and postdoc Jason Abercrombie)

Top 10 characteristics of a good mentor

Finding a good mentor in science can be difficult because the most important characteristics of a mentor have more to do with character and personal integrity than immediately apparent qualities such as verbally demonstrated knowledge and publication record. Graduate school can be an arduous path for not only your development as a scientist, but also for the refining of your interpersonal communication skills and character. If you pay attention, you will be constantly learning from people who have been around the block many more times than you and have learned and mastered, forgotten, or never learned some of the more important life lessons. When deciding on a lab to work in, it's a good idea to ask the current students and postdocs whether the professor demonstrates these qualities (good or bad).

1. **Understanding of individual strengths, weaknesses, and unique talents.** A good mentor knows that each graduate student has his or her own specialized tool belt of strengths that he or she can bring to the lab, while also recognizing weaknesses. Good mentors will help to identify those strengths and unique talents while also helping the student improve in areas where skills are lacking.
2. **Clarity.** It is important as a graduate student that you know early on exactly what is expected of you in terms of work hours, specific duties, and number of publications required for your successful completion of your degree. A good mentor will clearly communicate these expectations at the start of the graduate program. The mentor will help you create a reasonable timeline in order to sufficiently complete those goals.
3. **Good sense of humor.** A good sense of humor is critical for a mentor in science. A mentor that has a good sense of humor makes the workplace a more pleasant place to be and has better relationships with his or her students. Learning to develop a sense of humor if you don't have one is equally important as you interact and develop collaborations with other scientists.
4. **Approachability and a good listener.** When you consider the seemingly infinite number of questions you'll be asking your mentor during your course of study, a good mentor will always be approachable to help answer those questions and provide you with guidance. The better listener your major professor turns out to be, the more fruitful your research will become. If he or she truly listens to the problems you are having in your research, then the most effective solutions can be quickly attained, and everyone wins!

5. **Capacity for correction.** As a student of science, you are going to be wrong a lot! But whether mistakes are the result of poor experimental technique or creating interpersonal drama within the lab, a good mentor has the ability to recognize these errors and correct both bad science and bad behavior.

6. **Diplomacy.** A good mentor has to be diplomatic in order to effectively solve problems and deal with employees with conflicting personalities and behaviors. Diplomacy also is critical for fostering collaborations with other scientists. Diplomacy is an important life skill to develop over the course of your career as you encounter conflicts and disagreements.

7. **Organization.** If during your visit to a prospective professor's office, you feel like you've just entered "ground zero," you might want to reconsider. An organized person can foster organization and productivity. The messy person is usually a mess.

8. **Leadership and vision.** A good mentor must provide leadership for the lab. Leadership without micromanagement allows graduate students and postdocs to make their own way without excessive floundering. A good mentor also realizes that science is a process and that we "stand on the shoulders of giants" (Isaac Newton). Good mentors always see beyond the struggles of scientific discovery to the potential that science can benefit society. This vision enables the professor to see the value in their research.

9. **Forgiveness and patience.** The masters or PhD experience can be quite stressful and can result in harsh behavior by the student. A good mentor recognizes the potential in their students, and forgives them when stress brings out the worst. He or she understands that graduate training is not easy. The mentor can separate behavior from a person's integral being. Few things in life are more rewarding than the results of patience. A mentor who has a lot of patience has wisdom. A good mentor recognizes the inexperience of his or her students, and is patient during the course of their development as scientists. Patience and forgiveness go together.

10. **Communication.** A science professor is usually writing grants and manuscripts, and attending national and international scientific meetings. Therefore, he or she must be a good communicator. A proven track record of publications

in relatively high-impact journals and a high attendance record at scientific meetings is a good reflection of the professor's communication skills.

Top characteristics of a bad mentor

1. **Anger.** If a professor has a reputation of being easily angered, or is known as a "hothead," this can make your life difficult and create an anxiety-filled work environment. People burdened with unresolved anger issues generally don't make good mentors. Be careful not to become like this.
2. **Greed.** When research is driven by greed, and not science, bad things usually happen (e.g., falsification of data). By nature, science is an altruistic enterprise.
3. **Apathy.** When hard work and diligence goes unappreciated or unnoticed, it can be perceived as apathy or the expression of a professor's lack of interest in what their students are doing or struggling through during their research. A bad mentor is exclusively focused on his or her own pursuits and is disinterested in the struggles and problems of his or her students.
4. **Condescension.** If someone is being condescending, it gives the impression that an individual has an intrinsic feeling of higher self-worth more than someone else. This is a bad trait for anyone to express, especially a mentor. Condescending individuals cannot mentor students because their students will always feel a lack of respect and dignity.
5. **Arrogance.** There is a fine line between confidence and arrogance, and often, people in science cross that line routinely; especially apparent in the arena of battling egos. However, someone with a reputation of being arrogant does not make a good mentor. Nobody likes associating with arrogant people.
6. **Complaining.** Constant complaining, grumbling, passive-aggressive behavior, sour grapes, and the like are antithetical to being an effective mentor. If all you're doing is grumbling about how all the experiments are failing, why grants are being awarded to someone less worthy, blaming your circumstances on other people, and whining, then it is a reflection of your own poor performance.
7. **Use of fear or intimidation to produce results.** Although aggressive bosses may sometimes do well in science and lead very "productive" labs, they are not

going to express sympathy when you need to take some time off or have some life outside of the lab. During the process of completing a PhD, there will be many times when your research is consuming your life (and your thoughts) and having a PI breathing down your neck demanding more and more data is enough to make anyone miserable.

8. **Absenteeism.** The absentee mentor can't mentor. Some PIs take long and frequent vacations, sabbaticals, and any other excuse than to be at the university; they can become out of touch with students' programs. A derivation from the absentee mentor is the mentor who comes to work every day but takes seemingly never-ending coffee and lunch breaks. While networking is needed in science, hours upon hours with colleagues each day socializing sets a bad example for students.

9. **Workaholism.** The opposite of the absentee mentor would seem to be good, but in fact, the workaholic mentor can be worse than the absentee mentor. This mentor lives only for work and science and believes that you should be the same way. Not only is it unhealthy, but fewer scientists these days find it sustainable.

How to train your mentor

Let"s say that you want your mentor to be more helpful. Is it possible to improve your mentor's ability to mentor and therefore improve your own training? Let's say you want your mentor to be more responsive to your needs. It is hard enough to change our own behavior and personality traits, so don't even try to change your mentor's. Instead, let's assume that all mentors' innate desire is to put their best efforts toward great research. And by doing so, the mentor needs trainees to perform their best and get the best results. So, by that logic, one way to make a good mentor is to be a good student – see the Laws of Herman. I think that most faculty members instinctively want to "clone" themselves; i.e., produce a scholar who will follow in their own footsteps in the same field and be a faculty member at a research university. I know that's how I imagine students' destinies when they walk through my office door for the first time. I think, is this "the one"? The one who will do bigger and greater things in my field than me? Well, that approach is neither practical nor realistic, but one way to train your mentor is to allow the PI to think that the student wants to follow in the PI's footsteps. I'll admit this trick is somewhat disingenuous and I hesitate to mention it here because it could be construed as unethical, but the student really does want to learn everything the mentor knows and more, so this approach can be practical and effective. Students also have to be assertive to get good mentoring. For example, ask your mentor questions about the profession and the various options for your career path after you take the PhD. Maybe certain kinds of stress the mentor assumes is not for you, but there's no reason to lay all your cards on the table about career preferences and goals in every detail. The most important thing is to strive for excellence and that will seize the attention of your PI, giving you credibility and leverage to obtain what you need with regards to mentoring. As a mentor, I also appreciate students who keep me informed and want to dialogue about all their positive results in the lab. I appreciate the ones who ask me questions, come up with their own ideas and experiments and expect mentoring. Those things motivate me to be a more responsive mentor. In science, the shrinking violets are never rewarded. Don't be shy with your mentor and expect to be mentored!

Case study: choosing between two possible mentors

Joel has just finished his master's degree and is now looking at various universities and mentor candidates for his PhD. His master's degree was in the field of physiology and now he is looking toward adding mathematical and computational approaches to his repertoire. While Joel didn't publish any papers from his masters work from Comprehensive State U, he learned useful lab techniques, has high GRE scores, especially in the quantitative portion of the exam, and shows good promise as a researcher. His master's lab was small and in a department where the master's degree was the highest one offered. Joel was a teaching assistant and is comfortable teaching and working in a small lab.

He visits two professors at Big Private U, a school where his masters professor recommended for further study in computational biology. The first professor he visits, Professor Uno, has exactly the sort research program that Joel is interested in, but Uno's lab is small and with little funding. Professor Uno is friend of Joel's masters professor. Professor Uno says that Joel is a good candidate for a teaching assistantship, and already has a project exactly planned for a new student that would be perfect for him. He shows Joel an old grant proposal. Professor Uno says that while the proposal was declined for funding when it was submitted to the NSF a few years ago, and then again last year, that he still believes the project is a sound and solid contribution. He plans on submitting it again: "Third time's a charm," he says. Uno also presents Joel with a schedule of month-to-month milestones and says that he will help Joel to stay on task by working side-by-side with him each day. He will also need weekly reports from Joel. Professor Uno motto is, "Leave nothing to chance." Professor Uno has not published extensively, but his papers seem good – typically Uno is the first author, and proudly tells Joel that he is a perfectionist and will spend several weeks poring over each section of Joel's writing. Professor Uno expects each sentence to be perfect. His graduate students are typically coauthors. Joel likes the thought of the one-on-one attention. In addition, Professor Uno reminds Joel a lot of his masters professor, but with a better publication record. Joel's CSU professor didn't publish much and Joel was looking forward to really seeing how publishing is done. While Professor Uno does not show Joel his laboratory during his visit, he says that the computational facilities across campus are exceptional and that the campus supercomputer can be used for Joel's project. In addition, Professor Uno guarantees Joel will be able to finish his PhD dissertation in just three years and be directly competitive for a faculty position in the field. At the end of their meeting, Uno asks Joel not to talk with anyone about the project he has described that could lead to Joel's dissertation. Uno says that it is important to keep the project secret because someone else might steal the idea and get funding with it instead. Joel agrees to keep the project secret.

Joel then visits with a second faculty member (Professor Tua) during his day-long tour of BPU, but she does not have as much time to spend with Joel as Uno did. She introduces him to several graduate students and postdocs in his lab instead of talking with Joel for his entire visit. Professor Tua has several

NIH grants and is working in both physiology and mathematical modeling. The project she can offer Joel is part of one grant, in which the lead work will be done by a postdoc, Dr. Quattro, who has extensive experience in molecular modeling. Dr. Quattro seems to be a nice person who is very focused on science. Both Quattro and Tua say that Joel should get between three and four first author papers if he works hard and shows initiative. But neither Professor Tua nor Dr. Quattro is quite as warm as Professor Uno, but they both are well-respected and liked by the other people Joel meets in the lab. Dr. Quattro also shows Joel the computer cluster where most of the modeling work is done and, in disagreement from what Professor Uno said, Quattro claims that the supercomputer across campus is not readily accessible and would also not be needed for any project that Joel would likely do. This apparent disagreement about equipment accessibility and need is somewhat confusing to Joel and he thinks that not getting the opportunity to use the supercomputer would be a disappointment. Professor Tua's lab is also quite larger and more productive than the small lab Joel did his masters work in, but it is also more intimidating. Professor Tua says that Joel would be on a research assistantship and would work closely with Dr. Quattro. While Joel wouldn't be working directly with Professor Tua in any aspects of the research, she has an open door policy where people can walk in with any questions. She also says that it might take Joel longer than three years to do his PhD research. She notes that the time for data collection and writing the papers for publication is sometimes lengthy and unpredictable. She did say, however, that her 20 past PhD students have taken between three and five years to complete their PhDs between three and five years, but that she cannot guarantee that Joel will ever complete the PhD, because much of the success or failure is up to each individual student.

Discussion questions

1. Joel is comfortable with Professor Uno's hands-on approach, which is more like his previous mentor. How important is this feeling of comfort and familiarity in making his decision?

2. How important is it to work on a funded grant or at least, a line of research that seems fundable?

3. What should Joel think about Professor Uno's desire to keep the project secret? Do ideas often get stolen in science?

4. Should Joel worry much about the prospect of working more with Dr. Quattro instead of Professor Tua in his PhD? In contrast, he'd be working very closely with Professor Uno should he choose his lab.

5. How should Joel feel about the different information he receives about the supercomputer? How should that affect his decision?

6. How should Joel feel about Uno's guarantee that Joel would complete his PhD in three years versus the absence of any guarantees with Tua?

7. Which lab should Joel join and why?

Case study: choosing the right research project: the new graduate student's dilemma

Case study courtesy of Jonathan Willis

Karen Sparks joined Dr. Amie Leavens' lab as a new masters student in a marine science department. Karen aspires to eventually take a PhD in marine biology and wants to choose a research topic to eventually position her for a tenure track position at a research university. Karen has just started learning molecular biology.

Dr. Leavens has recently received tenure and won two grants. The first is an applied project on environmental toxicology of copepods and the second is more of a basic study of the molecular interactions within the copepod and how gene expression changes because of environmental changes. At the end of Karen's first semester, Dr. Leavens approaches her wanting to know which of the projects she prefers to work on and asks her to set up her graduate committee.

Dr. Leavens believes that a masters study should be very focused and straightforward. She suggests that Karen could simply execute a portion of the more applied toxicology grant as written. The molecular study is far more in depth, more uncertain as to the results, and would involve more time and some additional experiments that would require some methodology to be worked out. Dr. Leavens also mentions that she thinks that perhaps one large or two smaller papers will come from the toxicology work, but at least two- and possibly three bigger and more important papers could result from the molecular work.

Dr. Leavens has funding for two years of toxicology studies and three years for the molecular research. She feels strongly that more funding could be attained for future molecular work based on very early preliminary data. There is a post-doc in Dr. Leavens' lab who will perform molecular biology on a different goal of the grant. If Karen chooses the molecular route, she will work closely with the postdoc. If she chooses the toxicology project, it would be solely up to Karen to perform the research. Dr. Leavens tells Karen that based on her background she feels confident that she can achieve goals pursuing either project, but the molecular project will require more creative experimental design and more work than the other. She tells Karen that there is more certainty of completing the toxicology research in two years but cannot predict what will be the actual timeline for the molecular study given the circumstances.

Discussion questions

1. What factors should Karen weigh into making her final decision for her graduate program?

2. For the toxicology project, she will be a more independent worker, but not play a creative role in experimental methods, since those have been finalized in the grant proposal. How much should that matter to Karen?

3. A tradition doctorate program in biology ranges from three to five years, depending whether a master's degree is earned, how fast

the research becomes productive, and other factors. If Karen pursues the doctorate with the molecular study, she is not guaranteed funding after the third year. How important is not having guaranteed funding after three years to the student?

Summary

✓ Due diligence in picking a lab and a faculty mentor is crucial as a springboard to a successful career in science.

✓ It is important to like and respect your mentor; therefore, it is important to search for likeable and productive mentor candidates who possess traits that command respect.

✓ Don't be fooled by words – check out potential mentor's claims by actions and their research record.

✓ Trainees should be vigilant to avoid finding themselves in a toxic environment.

✓ Graduate student and postdoc trainees must be assertive to receive the mentoring they deserve.

Chapter 5

Becoming the Perfect Mentor

ABOUT THIS CHAPTER

- Good mentoring of junior scientists helps them to become successful senior scientists.
- Imparting skills to foster independence is the goal of mentoring the next generation of scientists.
- Teaching trainees how to publish results and acquire funding are two vital skills needed for success.

After all the schooling and postdoctoral work is completed and, when you have your own lab, what kind of mentor will you be? The best mentors allow their students and employees to fulfill their own personal and professional goals while still advancing science. One of the first questions I commonly ask the visiting student or postdoc candidate is, "what do you want to be when you grow up?" The answer reveals the person's career goals. In addition, how they approach this question can be telling. It is rare that anyone has long range goals clearly delineated, and therefore, mentors have a huge opportunity to help trainees clarify fuzzy goals and to develop milestones to fulfill goals. As mentees mature and grow as scientists, the mentor has the privilege of helping them find and follow their dreams. Mentoring comes in stages, and the kind of mentor you are early in your career might change as your career advances, and that's normal. It seems to me that as I get older, enabling others' success becomes an important way for me to be successful in science. Just as mentoring styles and goals change during career stages, each trainee has individual needs. Some scientists-in-training need significant guidance, especially in

---

Research Ethics for Scientists: A Companion for Students, Second Edition. C. Neal Stewart, Jr.
© 2023 John Wiley & Sons Ltd. Published 2023 by John Wiley & Sons Ltd.
Companion website: www.wiley.com/go/stewart/researchethics2

their early training. Others simply need to be pointed in the right direction and have their paths cleared of obstacles that stand in the way of success. A good mentor will be able to detect how they can best help each trainee in specific ways, and then understand how to meet the professional needs of each person.

Building your team: when to say "no"

Faculty members and staff scientists build a team of researchers who work independently and together to accomplish research. Some decisions on who to hire are easy. Here are some "types": 1) The superstar: the bright, eager, and diligent students and postdocs who are a great fit for research projects and are excited about being a scientist. Superstars are relatively rare. Hire them, reward them, and be happy when they move on to start their own labs (see the Bob Langer interview at the end of this chapter). 2) The capable: these are the potential lab members who are good hires. The capable have relevant experience, have a decent track record, and will be pleasant lab mates. Most labs are mostly composed of the capable. 3) The problem child: take whatever measures necessary to identify them before you hire (and whenever you sense that your interviewee is a problem child, just say "no" to hiring). The problem child may be charming during the interview and have one or more apparent strong suits. She will give you reasons to hire her. He may tug at your heartstrings. These may be people you think should have better track records and would if they'd had better mentors, parents, or circumstances growing up. Here are a few of things that may be "red flags". 1) No letters of reference from a key mentor/major professor. The problem child alienated their mentors by misbehaving in an earlier stop. 2) No strong first-authored publications or even worse, no publications. The problem child was incapable or too lazy to do the experiments in the right way and be a strong finisher. There will be no shortage of excuses and they will attempt to convince you that this time will be different. 3) There is no really good professional reason why the child should be hired. When viewed objectively, there won't be any compelling reason for bringing the applicant on board. Yet, sometimes, against all odds, someone on your team really wants to hire the problem child in spite of the red flags. Maybe a key member of your staff wants to "give this person a chance" when most of your team is disinterested or negative. Just say "no." If you, "the boss," have a particularly bad feeling about hiring someone, just say "no." What if you end up saying "yes" to hiring and then regret it early-on, when the negatives of the problem child are manifested. If your organization has a probationary period, then you're in luck and you can control the level of damage to your group by firing the person: good for you. You will have [mostly] dodged the bullet. But if you hire and don't fire, the problem child will disrupt your lab, bully your staff, create chaos, and in general, make your life miserable until you are able to convince the person to move on or fire the problem child for cause. And even after the separation, and all you have left is a few tattered lab books to show for it, the problem will persist long after the child is gone. People in your group (and you) will mostly recover from the damage, but the negative effects will last a long time. So heed warnings and just say "no" to hiring the problem child. You owe it to yourself and your productive students and scientists on your staff.

Grants and contracts are a prerequisite to productive science

The professional aspects of becoming a good scientist always return back to productivity: publications as well as grants to fund the research that lead to publications. If a graduate student or postdoc wants only to teach, then grantsmanship would appear to be less important, but what if the student has a change of heart and also wants to pursue research? Unless they go into industry, raising funds for research is an important skill. I guarantee that a teaching college will find evidence of grant-getting attractive, because more and more administrators in teaching colleges view research and teaching grants as vital to expanding and improving their programs. Therefore, mentoring is multidimensional, even in the simplest of desired outputs.

I argue here that becoming grant-savvy and a grant-getter has value no matter the career goal. I frequently ask graduate students and postdocs to write portions of grant proposals where I am the PI. After that, I may request that they craft their own proposals for either me to submit or, if they wish, they can serve as PIs and submit them on their own if the university allows it. It is often best if I serve as the PI on the grant since most universities do not allow graduate students, and sometimes do not allow postdocs, to be the PI on a regular, full-sized, grant. If I am the PI, ethics prescribe that I confer control of the science to the postdocs who wrote the proposal, and monitor, then transfer the grant to them outright should they transition to positions appropriate for them to be the PI for the grant, such as when someone becomes an assistant professor. I can also help them become familiar with the budgetary and paperwork aspects of proposals and grants after they are funded. As a mentor, I can also help them manage the people aspects of research.

Grant proposals are not easily funded, in part, because there is a learning curve in writing a fundable plan. The research also must be modern and thorough. For most scientists, it usually takes trial and error before winning. Even for experienced PIs, there is difficulty in acquiring funding, which is always statistically, improbable. But the probabilities fall to zero if a proposal goes unwritten and unsubmitted. I like to think of practicing writing grant proposals to be a relatively low risk activity for graduate students and postdocs since they typically don't absolutely have to win their own funding to succeed as scientists at very early career stages. In most cases, they are funded for their existing positions in another grant. Therefore, it is a win–win situation if their proposals are funded but not absolutely necessary to be deemed successful as postdocs, for example. This situation changes drastically when the tenure clock begins to tick. An assistant professor needs to be successful in the grants world to earn tenure at many institutions. I prefer for my postdoc trainees to already have "paid their dues" in grant-rejection-land prior to when it actually counts as their tenure clocks tick. Naturally, they will still experience grant proposal rejection as assistant professors, but chances are they'll also experience a higher frequency of success.

Judge yourself

✓ Do you enjoy planning experiments and asking for resources to do research? Do you enjoy teaching these skills to trainees?

✓ How much do you believe in your abilities to find funding for your research and teach others how to do the same? Can you imbue confidence to others?

✓ Are you a trusting person? Can you rely on others to succeed? Are you devastated if your trainees fail? How do you deal with success? And cope with failure?

Publications are the fruit of research

One huge role of mentors is to enable research to be published. The primary route is through the publication of theses, dissertations, and postdoc research in the form of journal articles. This means allowing and expecting trainees to take control of a study and the resulting manuscript: own it, write it, be the first author. It also means encouraging teamwork and coauthorship. It is bad mentoring to allow important research that is part of a thesis to languish unpublished. But sometimes this still happens and it does not make me happy. The best scenario is for students to author manuscripts for publication in journals before they write their dissertations. After a student leaves the lab, a professor's leverage is lower and passive trainees might never write-up their results and publish. On rare occasions papers go unwritten, unless the professor does it; students sometimes leave, moving on to other pursuits, and forget about their unwritten manuscripts. While professor-written, erstwhile student papers might help science (better than the data going unpublished), this practice defeats a major purpose of mentoring: trainees should master all aspects of manuscript preparation, peer-review, and revision. Under the umbrella of the mentor, this mastery optimally occurs when there is the option to fail with little risk or penalty; this situation changes when the tenure clock is ticking. Trainees must gain skills of scientific independence if they are to become fruitful contributing senior scientists.

My first priority as a professor is to encourage manuscripts to be drafted, vetted by all the co-authors, finalized, and then submitted as quickly as possible. I don't understand why a professor will sit on a manuscript for weeks and months, but I see this happening around me all too often. I hear their students and postdocs complain bitterly of this situation. I don't think their mentors appreciate the depth of their trainees loathing when their PIs procrastinate and badly manage research. I can understand some delay in submitting a paper if a grant proposal is taking precedence, but students and postdocs need timely publications for them to realize their professional dreams, which typically includes graduation and career advancement.

On a personal level

Good mentors also realize that their mentees are more than science robots. They are people with feelings and personal goals and dreams. I've listened to students talk about their friendships, love lives, parental aspirations, health problems, religious beliefs, politics, philosophies, and exercise regimes. I have given advice on what classes to take, what cars to buy (or avoid), who is a trustworthy collaborator, and the difference between ethics and ethnics. I've traded stories and had arguments about which barbeque, beer, and football teams are the best, renting vs. buying, boats, musical instruments, table tennis and why christopherwalkeniswatchingyoupee.com is both kind of creepy and cool at the same time. The best mentors continually listen to their trainees and give advice when needed, and not the other way around. Effective mentors should sense when their trainees aspirations change. Personal situations affect professional performance, therefore mentoring must be holistic in nature. That said, trainees' personal lives and boundaries must be respected. No one likes a nosy boss, but everyone wants someone to care about them on a personal level. There is a fine line to be observed here. And in no circumstances do trainees want to hear about the boss's personal troubles. It is ok for mentors to lament about their recent manuscript being rejected but is not ok to talk about their domestic arguments.

Graduate students and postdocs need freedom do their desired experiments and follow their noses that will lead to good publications. I stopped doing lab work more than two decades ago. Therefore, since my two hands are no longer performing experiments, the science that comes out of my lab is the only science I can claim, albeit vicariously. I have no backup plan for collecting data and generating papers – I'm completely dependent on my trainees to succeed. The mentor is more like a coach than a player, so let's end this section with quotes from the late basketball coach John Wooden. "A coach is someone who can give correction without causing resentment." And, "You can't live a perfect day without doing something for someone who will never be able to repay you." These words paint the picture of a perfect mentor.

Judge yourself

✓ How much do you care about people? Specifically, how much do you care about people on your team?

✓ Are you empathetic? Are you aware of lines that you should not cross that would otherwise make people intimidated or uncomfortable?

Common and predictable mistakes scientist make at key stages in their training and careers and how being a good mentor can make improvements

Graduate students

Many graduate students don't immediately make the transition from being an undergraduate and having knowledge spoon-fed to them to being sufficiently aggressive to obtain knowledge for themselves. The US system, which requires certain additional coursework during the masters and PhD degrees, doesn't help this situation much, in my opinion. Requiring formal courses is not all bad, but if graduate students are taught in the style as when they were undergraduates, this passive learning style is reinforced. If students are at their desks "studying" more than they are in the research lab doing experiments, they are not becoming independent scientists. Also, nearly all students make the mistake of not doing enough independent reading in their field. Likewise, they also seldom write enough. And when they do write, it is often not in a style that is conducive to conversion to publishable journal articles. Many science students are not good writers, but if they don't become comfortable writing and speaking in scientific English, their careers will never be fully launched or their scientific potential realized.

Graduate students can become mentors to undergraduate students performing research in the lab and also to entering graduate students. This mentoring not only helps the trainees, but also aids in graduate students making the transition to becoming a professional scientist. It is well-known that teaching helps the teacher gain a deeper understanding of topic taught. Indeed, students respond well to mentors who are often not far beyond their years and experiences.

Postdoctoral fellows

The common early mistake of a postdoc is to dwell a bit too long and too seriously about the glories of finally obtaining the PhD and having the title of doctor. Postdocs don't realize soon enough that they are really just getting started in research and have a lot to learn. Suddenly growing a big ego, strutting around the lab and barking out orders, is not the way to engender learning or cooperation from those around you. Another common problem results when postdocs did not grow sufficiently independent as researchers in their studentships and are not quite ready to be independent postdocs. I've heard my lab manager utter the phrase, "you're a postdoc now" in an attempt to inspire independence. There are many times when this phrase (which can be sung to the tune "You're in the Army Now") doesn't translate to instant independence, and maybe for good reason. When postdocs change fields and learn several new topics while simultaneously getting used to new surroundings, it might take some time for them to feel independent and productive. Nonetheless, postdocs must challenge themselves to gain a sense of independence and competence, because that is what sorts out effective scientists from the crowd. At the end of the postdoctoral period, scientists should have had the opportunity to learn from their own mistakes in mentoring, grant proposal writing, and publications so they don't have to make the same mistakes during their assistant professor years while their tenure clock is ticking.

They can help themselves by trying new tasks, including the submentoring of graduate students and undergrads – really taking charge of a project. Having recently completed dissertations, preliminary exams, and the angst of getting their degrees, postdocs can be of great comfort to the students they are mentoring. Mentoring also helps postdocs see beyond themselves and toward service to their labmates and greater science.

Assistant professors

Some assistant professors who have not had sufficient and effective postdoctoral training make predictable mistakes in writing grant proposals and mentoring. Being able to write effective grant proposals is crucial to getting tenure in most universities. The ability of assistant professors to win grants is actually more important than their doing experiments themselves in the lab. And since assistant professors might be increasingly absent from the lab, they must learn to recruit and retain graduate students and other trainees.

In fact, mentoring mistakes (beware of being a new hire's first grad student!) are bound to occur as the assistant professor is learning the ropes of mentoring. A big mistake made just after recruiting that first-graduate student is not letting go of the research reins. As we know, graduate students must own their own projects. Good mentors don't micromanage. They also are not too buddy-buddy with their trainees. The mentor is a faculty member now, not a student's peer. Collegiality is good. Sleeping with students is bad. Another mistake assistant professors often make is not having a balanced life. By worrying so much about making tenure, they don't cultivate interests outside of work and can actually become less effective and tenurable than if they pursued some personal activities outside of work. Other mistakes include teaching too many classes, taking too much time to prepare for teaching, and engaging in excessive university and professional service. In almost all universities, tenure is awarded primarily on research productivity (i.e., peer-reviewed papers and how much they are cited) as well as their grants.

Associate professors

Associate professors shouldn't take long naps and vacations after they are awarded tenure. Pausing research for even a short period of time results in a higher probability the research program will cease to be competent and competitive. When it ceases to be a fun challenge, it is best to leave research. One mistake that associate professors make, especially if they decrease their research productivity, is to start down the vortex of too much university and professional "service" and teaching, and not enough recruiting and mentoring of trainees. Feeling guilty about the lack of research productivity and also experiencing some immediate rewards from running big professional society-, department-, and university committees, serving on the faculty senate, and doing more teaching, the associate professor is now poised to be even less competitive in research and less likely to be an attractive mentor for graduate students and postdocs. Like tenure, being promoted to full professor is usually based upon research production and whether the candidate has an international reputation in his or her field. Therefore, serving in more editorial roles, organizing conferences, serving on grant panels, and research-oriented service are more effective paths to becoming a full professor than chairing a big faculty senate committee (leave that to the full professors who are close to the end of their careers).

Full professors

The proverbial "deadwood" at universities almost always is composed of full professors waiting to retire (albeit sometimes waiting decades!). With post-tenure review becoming more common, tenure-for-life is no longer a death-and-taxes certainty in academia. It seems to me that if someone chose to pursue science because of the intense interest in research, then why should that necessarily change with age? Some full professors lose the competitive edge and do not stay current in the field. It does not take long for science to accelerate past static competence and comfort into a realm of unfamiliarity. In finding that they are no longer competitive for grants, full professors sometimes quit doing research altogether. Sure, increased university service is rewarding and someone has to do it, but keeping current in science is more fun. And science is more fun than coming in late, going home early, and drinking coffee and shooting the breeze with the like-minded deadwood in between. In addition, the best full professors are efficient enough to have fun at more than one thing at a time.

It is good for the full professor with tenure to remember the fun of doing science. Sabbatical programs, outside training, and rekindling research in hands-on laboratory experiments can be helpful to jump-start research. Doing research in a very different field from the one in which a faculty member gained a reputation can also be scintillating. It is a big mistake to become a member of the deadwood coffee club. As the reticently retired American football coach Bobby Bowden (he retired at age 80) said, "After you retire, there is only one big event left." In this stage of their careers, full professors should now know enough to be expert mentors. Experience is a terrible thing to waste. Science should benefit the most from full professors sharing knowledge with their younger trainees. Full professors should be inspirational. See the Box on Bob Langer for an excellent example.

Mentorship of many trainees: the story of Robert Langer

Dr. Robert Langer, Institute Professor, Massachusetts Institute of Technology.

Source: Science History Institute/Wikimedia Commons.

Robert (Bob) Langer has a gigantic lab in Chemical Engineering at MIT. Helen Pearson, a writer for *Nature* followed him around one day to learn why and how he is so productive (Pearson 2009). As of 2022, Dr. Langer is one of 12 Institute Professors at MIT, the highest honor that MIT bestows, has published over 1500 papers, is the inventor of over 1400 issued or patents pending. He is the most cited engineer in history and has won over 220 major awards. He is one of few people ever to be elected as a member to all three National Academies (Engineering, Medicine, and Sciences). Given his extraordinary productivity and genius, I was even more impressed by why he does what he does: he says wants to help people and make them happy – so he said in the *Nature* article. Therefore, it seemed to me that Bob Langer might be the perfect person for me to ask a few questions about mentoring. Sure enough, he answered my request email after just a few minutes and agreed to an interview for the first edition of this book.

Neal: I read in *Nature* that you have 80 or so people in your lab at MIT, which would make it the largest research lab at, arguably, the best scientific research institute in the world – maybe the largest research lab in the world. The size seems overwhelming to me. Could you tell us something about the breakdown – the number of post-docs, graduate students, etc., and what is the most over-whelming part of managing such a big group?

Bob: Currently, I have probably around 25 graduate students, maybe 45 visiting scientists and postdoctoral fellows, 8 technicians, and 5 office staff. At any time there are 30–50 undergraduates in the lab doing research (called UROPS at MIT). It's impossible to give you an exact breakdown. The most overwhelming part? You know, it's funny – I don't really find it that hard. Growing over time might be hard, I suppose, but maintaining a steady state is not that hard. You just develop a style that works. I have a really good office staff and there are senior postdocs – it just works itself.

Neal: You must be a good manager of your time and resources. What is your "secret"? What advice could you give assistant professors embarking on a career in academic research science?

Bob: I think I am good at delegating. I don't know that I have any secret. Obviously. I work hard ... lots of people work hard. I've been a pretty good delegator. I don't think it's important that you have a big or small group. The advice I always give people is that it is important to work on projects that you really believe in; projects that you feel will make a difference. It's important to take risks and work on really important high impact projects. Be good to your students and postdocs – help them in any way that you can. It's really important to obtain solid funding. These are some of the priorities that I tell people to do.

Neal: What is your mentoring style? How does mentoring differ among people in your lab? That is, do you mentor graduate students differently from undergrads? Postdocs differently from grad students?

Bob: To me, I consider that people are at different stages of life. Until the time you do research, you spend your life answering questions people ask of you. If you're in high school or in college courses you're judged by how well you do on tests. Even as an undergraduate and sometimes graduate student working on research projects that other people come up with – you are judged on how well you do on them. My goals for undergraduates are to learn good lab skills and get excited about research. For grad students – I want them to begin to make the transition from giving answers to asking questions. And for postdocs, even more of that. Below is my 2009 interview with him.

Neal: So it seems that you encourage your students and postdocs to "own" their projects. Do they take the projects with them when they leave your lab, or do you keep part of the projects? What is your policy?

Bob: I do tell people it is fine to take their projects. However, it important for their sakes that if they take aspects of a project that they do something different with it. If the NIH or their new faculty think they're doing something that just duplicates their postdoc project, they won't look favorably on them. I think it's best for people to do things that build on what they've done. But if someone want so take their project, we have so many things going on that it's ok. But I want to encourage people to think big and build on what they've done.

Neal: Do you have an open door policy? If not, how does your staff access you?

Bob: I used to have an open door policy, but now I have an open book policy. People just put their names in my book or tell my secretary, or they also email me and I try to get back to them quickly.

Neal: I did notice that you are a rapid email responder. How do you manage all of the information you are barraged by? From emails to the literature and things in between?

Bob: I don't like things to build up and I like really try to finish everything everyday so they don't build up. Otherwise I get stressed and if I get things done I don't feel stressed. I feel like people appreciate it when I get back to them on email and that means a lot to me. Like the article in *Nature* said, I like to make people happy. I'm the kind of worker that'll keep going at it till I've finished the things I need to do. For example I'll do email while I'm walking. I do answer a lot of email.

Neal: What do you think is the most common mentoring mistake that faculty members make? The most damaging?

Bob: I think that faculty members have to walk the line. Some faculty members are too controlling from my stand-point, and maybe others are not [controlling] enough. Sometimes they tell people too much of what to do and then students don't get excited or don't learn the keys of what research really is or how exciting research can be. I think it's important to give students a certain amount of rope and to show them you're always there for them. I think the biggest mistake is oversupervising.

Neal: On the other hand, what is the best thing a mentor can do for his postdocs? His graduate students?

Bob: The best thing that you can do is to help them get really great ideas, and how to do research that has a huge impact – to show them the excitement of research.

Research can really be a wonderful thing. You want them to have good credentials, good papers, and good grants.

Neal: In an earlier answer, you used the phrase "walk the line." I think true artists and successful scientists are highly adaptable – that if Johnny Cash were starting out in music today he'd be poplar in some genre of music. And if past Nobel prize winners were starting out in science today that they'd contribute in modern science too. What is the mentor's role in nurturing adaptability and "genius," or do you think these are characteristics people are born with?

Bob: This a really interesting question. I think it is some combination of people being born with talents and other things. But I think that mentoring plays a big role. I also think that circumstances can play a big role. Maybe you've read Malcolm Gladwell's book "Outliers." Circumstances about where you are in history can play a big role. Mentors can play a big role. For me, I had Judah Folkman as my mentor, and that played a big role. Mentoring certainly helped me. I don't think there's any accident that there's been a couple hundred people from my group who have gone on to be professors and another couple hundred people from my lab who are now entrepreneurs. I saw success from him [Folkman] and he was a good mentor for me. And I've liked to see people from my lab do well. They're like family to me.

Neal: I'd appreciate any other words of wisdom you might have for young scientists starting out in research.

Bob: Gee, I'm not sure. I think goes back to what you asked before. I think that it's really important to pick challenging problems that have high impact and then be sure to raise enough money.

In 2022 when I was working on this second edition of the book, I reached out again to Bob Langer for an update. Below are my questions and his answers.

Neal: The last time I contacted you (2009) you had a giant research lab group. Is it still giant? How has your mentoring style changed over the years?

Bob: It is. It's over 100 now. My mentoring style is pretty much the same.

Neal: You're a co-founder of Moderna. Do any of your trainees participate in the company? How does it feel to have made such a positive contribution such as the mRNA vaccine development in protecting people against COVID-19?

Bob: Yes, I've had a number of trainees work there. It's been an honor to have been involved with Moderna and to have been a part of what they've done for the world.

Neal: How did the pandemic affect the progress of research at MIT? How did your lab cope with the challenging circumstances?

Bob: It slowed progress but not that much. We did more zooms but a lot of what we do is very relevant to the COVID pandemic since it involved novel vaccine development. We didn't let up.

Neal: How concerned are you about anti-science sentiments that seem to have grown during the pandemic? How do you advise your trainees to prepare for their futures these days?

Bob: I'm concerned about it. It's disappointing. On my trainees – I advise them the way I always have – try to do projects that will be great science, that will have a big impact and treat people well.

Neal: Do you have any other words of wisdom for new scientists just starting their careers?

Bob: I think it's important to think of innovative ways to raise funds. Not just grants but also patents, foundations, etc.

Mentoring advice from a Bob Langer trainee: Daniel Anderson

Dr. Daniel G. Anderson, Professor, Massachusetts Institute of Technology.
Source: Massachusetts Institute of Technology.

Hearing Bob Langer discuss mentoring had a profound impact on me. Bob's mentoring philosophy provided confirmation that I was doing certain things right and challenged me to make improvements. Playing the skeptic led me to check out Bob Langer's story, so I also interviewed one of his long-term trainees, Daniel Anderson, who has been a research associate in the Langer group for about 10 years and who is now an associate professor at MIT. During this time, Daniel has sort of built a lab within a lab and tells me about his own mentoring style and what life is like in the Langer Lab at MIT. Below is my 2009 interview with Dan.

Neal: Tell me about yourself. What is your background? How did you find yourself in Bob Langer's lab?

Daniel: My PhD research was in the field of molecular genetics studying enzymes involved in DNA repair. My degree came from UC-Davis. I came to MIT as a postdoc and had the chance to stay and build a subgroup. I wanted to move toward more applied research and gene therapy. My subgroup consists of about 40 people: mainly postdocs doing research in the area drug delivery.

Neal: When you interviewed with Bob prior to joining his lab, do you remember any outstanding moment or event that shaped your decision to join his group?

Daniel: When you visit a lab, you want to get a sense of the impact of the place and people. You ask yourself, "Is this where you want to be to get stuff done?" You want to get a sense from other people in the lab as to whether they are happy and productive. On the other hand, you also want to sense if people are moody and if there are challenges to being productive. I think that if people are happy, that more work is accomplished, and so you screen for hard-working people who can work together as a team. So, when I visited at MIT and had a conversation with Bob, he offered me a postdoc position. I was struck with the impact that his lab had on science. It was clear to me that the work that was getting done had a high impact. Although I knew about the impact before my visit, during my visit I got a more tangible connection.

Neal: What do you think is the most common mentoring mistake that faculty members make? The most damaging?

Daniel: The one that sticks out to me is when people are jerks; when they treat their lab people like dirt. That one might not be the most common, but it hurts people and productivity the most. You need to be sensitive to people. Some mentors want to motivate people to getting every drop of productivity out they can, but it seems to me that by spending more time calming people down, they can get motivated. Mentors should motivate by the impact the research can have. But you have to be sensitive to people and their needs. I'm more interested in people being happy and productive. Typically, if they are happy, they'll be productive. We want people who are self-motivated.

Neal: What is the best thing a mentor can do for his post-docs? His graduate students?

Daniel: Spend time with them and think about what their projects are and help them. Mentors should read what their trainees write and help them.

Neal: What have you learned from Bob about mentoring, science, or life in general that you would like to share.

Daniel: In running a big lab, you have to have enough money to do it. Resources go beyond salary that is needed to do experiments. I've learned that the social challenges of dealing with large group of people is sometimes more difficult than the science; good interactions among people keep it healthy. Work hard and play hard. Good people come to us … finding money is a lot of work.

Just as I followed up with Bob Langer in 2022, I also checked in with Dan to see what changes he'd seen over the past dozen years and any new advice he might have for the readers of this book. Below is the text from my latest interview with Dan.

Neal: Back in 2009 when I contacted you and Bob Langer, you kind of ran a lab within Bob's lab. I see now you are a full professor and in another department at MIT. Are you still working closely with Bob? How has your research direction changed over the past dozen years or so?

Daniel: Bob and I still collaborate, but we're in different departments. I now have projects independent of Bob. I'm a full professor in the Department of Chemical Engineering and Institute for Medical Engineering and Science. My research has expanded to go in other directions.

Neal: One thing you said in 2009 that resonated with me was about the mentoring mistake of mentors. You said that some mentors "are jerks ... and treat their lab people like dirt." I've never understood why mentors would do that, yet they do.

Daniel: Ha ... I guess I could have put that another way, but it's true. A lot of employers only think about what their students and postdocs can do for them. Good mentors help their employees to grow. They spend time with them. Part of this is educational and spending time teaching them how to navigate science and also providing opportunities. Good mentors help their employees in ways that may not necessarily help the mentor's career: altruism is a mark of great mentors.

Neal: Addressing bullying in academia has come a long way in the past decade, but seems as if it has a long way to go. What are your thoughts on this topic?

Daniel: Bullying: I think about what children go through. But there has been a culture change – not only in science, but in other workplaces. It seems that people use to be more combative, but now they're more supportive. So that's a good change. I remember many years ago being in a seminar where the major professor was sitting on the front row where his student was giving a talk. The professor was cutting the student down in front of the department ... really wasn't being constructive, but combative. That kind of thing doesn't seem to happen these days.

Neal: Another thing you said in 2009 was that "good people come to us ... finding money is hard." Is that still the case?

Daniel: We've been fortunate in raising a lot of funds through writing proposals and talking with companies. We're also fortunate in getting a good people. You have to spend a lot of time to get great workers. But you know none of it is easy ... you just have to do it.

Neal: How have times changed in science with COVID and all?

Daniel: COVID was hard. People couldn't get in the lab and had limited abilities to do bench science. We had to find new ways to communicate and do research. There were no undergrads in the lab. But faculty still have the same job of doing research, teaching, and service.

Neal: Do you have any last words of advice for young scientists you can share here?

Daniel: Work with good people and do projects you're passionate about.

Case study: the case of the missing mentor

PhD student Mitch Mitchell is a first-year student in the lab of Dr. Thomas DeBague, which focuses on applied microbial biochemistry. The particular subject area of the lab has several practical applications and plenty of career opportunities for PhD-level scientists. Mitch enjoys this aspect of DeBague's expertise and initially viewed working in his lab as a great opportunity, since it is one of the few labs in this particular area of research. Mitch's main problem is difficulty in scheduling meaningful mentoring time with his major professor. Meetings with Dr. DeBague don't occur for several reasons. The first is Dr. DeBague's heavy travel schedule; second, poor health; and third, when he is in the building, it seems that he is either drinking coffee and eating donuts in the breakroom or sequestered away in his closed office door. A secondary, but related problem is that Dr. DeBague is not quick to respond to Mitch's emails or read and comment on Mitch's writing. Often, Mitch doesn't know if DeBague is in the office or even alive.

Another problem is that Mitch is the only graduate student in the lab and feels lonely. In fact, Mitch has few friends or social life and wants only to do science. He feels like he gets inadequate basic technical training in necessary biochemistry techniques, and virtually no mentoring. He wants a lot more direction for his PhD program. He did somewhat similar techniques for his master's degree, in which he was mentored by a young assistant professor, Dr. Fabé Energé, one of Dr. DeBague's first graduate students when Dr. DeBague was an assistant professor himself. In Mitch's master's lab, he got accustomed to an atmosphere of high levels of attention, and lots of day-to-day instructions and directions. Dr. Energé said that Dr. DeBague would be the perfect mentor for his PhD.

Mitch came to the conclusion that there must have been many changes in the intervening 16 years between Dr. Energé's and Mitch's tenure in Dr. DeBague's lab. When Mitch does manage to get Dr. DeBague's attention, DeBague compels Mitch to enter his office; DeBague then closes the door and subjects Mitch to seemingly endless stories about when Dr. DeBague was a graduate student and an assistant professor, how difficult his own graduate experience was, and how he has finally "made it" in science through his own ingenuity, energy, and hard work. Dr. DeBague is fond of saying, "I have tenure now so I can do whatever I want whenever I want to do it." After that cue, Mitch is typically then treated to travelogues about Dr. DeBague's latest vacation or conference. Mitch is very patient during this routine; perhaps too patient. Finally, whenever Mitch gets around to requesting professorial guidance, or asks about the proposal he gave him to read, Dr. DeBague finds something else that is more urgent. "I must now confer with my colleagues about an important project," DeBague will typically say. "Keep working hard and you'll be as successful as me."

Mitch does make modest progress in spite of being alone in the laboratory. He drafts his dissertation proposal early in his second semester as required. After two months, Dr. DeBague finally gives the proposal back to Mitch with very few comments. "Not a half bad proposal, if I say so myself," DeBague proclaims. Mitch's graduate committee generally lauds his efforts, but points out key technical flaws and scientific gaps in his proposal. They also point out that Mitch seems tentative during much of his proposal presentation. Mitch thanks his committee and vows to do a better job, but leaves the meeting feeling even more dejected and desperate.

Mitch feels that he is at crossroads as a graduate student. He has tried repeatedly to engage his major professor, but seems to be stuck in an endless loop that is never resolved or productive. What should he do?

Discussion questions

1. Should Mitch try a different strategy to get good mentoring from his major professor? If you think this is a good solution, how might he approach it? Is Mitch without fault in his situation?

2. What is the chance of success should Mitch decide to continue to go it alone as he has in hopes that Dr. DeBague is right? Do you believe the assertion that Mitch will be successful if he works hard? After all, isn't developing scientific independence a key outcome during the PhD process? What are the advantages and disadvantages of taking this approach? Is Mitch working toward independence?

3. Should Mitch leave the university or simply leave his department? Or should he stay in his department but switch labs? Or something else? He's taken a class with another professor who is engaged in research that is different from that what Mitch originally wanted to do, but Mitch has talked extensively about his problems with the other professor's graduate students and has seen merits in the new area; he thinks he might be able to cultivate interests in this area. The other students confide that since it is a larger lab that they aren't alone in science, but also complain that their professor is also very busy. However, they say they do see him and get directions and advice, but without the travelogue. Their professor also is responsive to email and reads and comments on their written projects – it only takes a few days to get back constructive comments. They encourage him to speak to their professor about switching labs. What are the advantages and disadvantages to switching labs and major professors? Would it place the alternative professor in a collegial bind to accept Mitch? Should Mitch change departments or universities to salvage his science career without Dr. DeBague?

4. Are there any other potential solutions?

Summary

✓ Learning to mentor is a process that takes courage to do well.
✓ Mentoring is dynamic and the roles and duties of mentors change throughout a person's career.
✓ Given the choice between being too controlling and less controlling, a mentor should choose the latter.
✓ A major goal in mentoring is to increase the independence of trainees.
✓ It is vital that mentors keep kindled their excitement for science.

Chapter 6

Research Misconduct: Fabricating Data and Falsification

ABOUT THIS CHAPTER

- The most blatant of research misconduct is the first F in FFP: fabrication of data, which is closely followed by the second F: falsification.
- The basis of sound science is sound and honest data.
- Scientists justify data fabrication with a myriad of reasons, none of which are valid.
- Image fraud – using figures inappropriately to deceive – is a growing problem in science that could be considered both fabrication and falsification.

Science rises and falls with data and their analysis. I believe that most scientists play it straight-up: they want to know and report the truth. They collect data from well-designed experiments and report the data that tell the whole truth. Not all experiments work because of technical problems and honest error; neither are all data created equally. Scientists are mainly an objective and honest group of people: highly educated and with an innate sense of research ethics. That said, research misconduct is rampant. It is visible in high-profile cases in high-profile journals. Honest researchers often become invisible to the public while the infamous cheater becomes notorious. Invariably, when such cases are made public, it is the result of discovery of misdeed when a reviewer, editor, or other scientists in the field discover that data are missing, wrong, or simply fabricated. Of course, honest mistakes happen in every field of endeavor, and science is no

Research Ethics for Scientists: A Companion for Students, Second Edition. C. Neal Stewart, Jr.
© 2023 John Wiley & Sons Ltd. Published 2023 by John Wiley & Sons Ltd.
Companion website: www.wiley.com/go/stewart/researchethics2

exception. Honest, even substantial, mistakes in a paper can result in the paper being retracted by the authors. Papers that are found with falsified or fabricated data are also typically retracted, but with more dire consequences that accompany findings of research misconduct. I know colleagues who believe that I and some other scientists over-emphasize the problem of research misconduct, but I believe it is actually underestimated. As a letter-writer to *The Scientist* pointed out: "If such major journals as *Nature* and *Science* could be fooled, imagine how many times they were fooled and fraud wasn't caught"? (Augenbraun 2008) And imagine how many editors of "ordinary" journals are tricked every year. After all, they are not staffed to the degree of *Nature* and *Science*. In addition, many editors are not vigilantly screening against FFP. In this chapter, we will examine a few real case examples of research misconduct with an eye toward prevention, so make sure to judge yourself throughout the chapter. Of special interest is the creation or alteration of illustrations or figures in papers and grant proposals using image manipulation software. Dos and don'ts will be emphasized when creating scientific images to avoid the appearance of fabrication and falsification.

Why cheat?

Successful scientists have the knack of coming up with great ideas. Great ideas are the foundation of scientific inquiry, but just having a great idea does not necessarily equal great science. Great science is defined as a great idea coupled with rapid and thorough discovery leading to diligent publication of the work that is regarded by and heuristic to other scientists. That is, the formula is great idea + rapid execution + sound data = great publication. Stating the formula is easier than realizing that great publication. Finding the big idea is tough enough. This is complicated by needing to be in the right place at the right time with the right people. Then the right people must execute science in such a way to get both the rapid collection of thorough and sound data. The summation of these leading to a break-through discovery and stellar paper is a rare occurrence indeed! I believe that the first part of the equation, a great idea, is often a much easier bird to catch than the second part – rapid and sound data. After all, I have come up with great ideas (or at least they seem great at the time) in the shower or at my desk, or walking through the woods. But to make science really happen is much more difficult and takes a lot of thought and work to acquire honest data. It requires other people too! A weak link in the team can prevent a scientifically sound paper from ever becoming reality.

First, you have to find the funds to do the research, and then you have to recruit the right people with the right skills to make the science happen. After these, the experiments have to "work" with none-to-minimal technical problems. And finally the data must be positive, i.e., demonstrate that the idea was sound and typically support (fail to reject, in science lingo) hypotheses tested with new knowledge being created. This concept of "positive data" is contentious in as much as "negative data" are typically not readily publishable in many fields of science. For example, let's say that I think that a certain technique using a novel chemical treatment will lead to the facile genetic engineering of an animal cell. I can test my hypothesis that adding the chemical will increase the transformation frequency. If it does, then my paper may be publishable, but if not (negative data), then almost no editors or peer reviewers will be very excited to publish the paper. Why? Because it adds no new knowledge to the field, except that the chemical didn't work. In the grand scheme of things, this attitude is correct. In another way, it is incorrect, because publishing an ineffective method and the sound data could save someone else much time and effort when they come up with the same "great" idea. The world of science rewards great ideas and speedy and elegant experiments showing those ideas have merit. Novelty is rewarded and learning nothing positive is not.

Speed is of the essence. So, scientists with great ideas and big labs have therefore solved a couple of crucial problems in executing experiments en route to the great paper. They probably already have some funds they can allot to small, but potentially impactful projects, and they probably also have accomplished lab staff (students, postdocs, etc.) who might be willing to participate in a potentially great discovery. Then the trick is to design the right experiments and hope that they work. When they do, it is magical and such experiences can define a scientist's career. The classic example of a great idea and speed is chronicled in James Watson's *The Double Helix* (1968) in which he tells the story of how he and Francis Crick pursued the elucidation of the structure of DNA, with special emphasis on their race against other scientists to make the discovery. Indeed, they used their and other people's data [although the surreptitious use of Rosalind

Franklin's data was ethically dubious] to synthesize the model that would prove to be the correct representation of reality. This led to a Nobel Prize for the achievement a few years later. Their competitors didn't win the Nobel. While there remains arguments about whether their process was entirely ethical and the subsequent award was completely justified, it turns out that the data and their model that fit the data were sound.

I believe that cheating in the form of data fabrication is often functionally envisaged as a way to replace rapid and thorough experimentation. There are two types of instances that this seems to occur, at least in many high profile cases. First, and curiously, many experiments do go well – they demonstrate that the idea is a great one and that the scientist does have insight into a problem. As a sidebar, it is important to note that cheaters are typically not incompetent scientists. But instances occur where key experiments that fail to yield the desired data. Therefore, there is the temptation to fill in the gap with fabricated data. I believe this is probably the most common kind of cheating of this sort. The scientist feels that the work needs to be published and published rapidly in a high-impact journal. Therefore, the argument the scientist may make internally is that the means justify the ends. When caught cheating, it is not uncommon for the scientists at fault to emphasize all that was right with a study and to apologize for "mistakes" made. They may also attempt to justify ill-gained shortcuts that led to "results" they published earlier, which was better than plodding along with subsequent experiments. The second type of cheating is more comprehensive data fabrication. This "Full Monty" of cheating seems to be less common, but it does occur. And when high-profile researchers commit Full Monty data fabrication they are certain to be discovered in their entire naked glory. Perhaps the most infamous case is Jan Hendrik Schön.

Before we get to the case of the Schön case, let's take a look at an example of what scientific journals expect in submitted papers. To paraphrase Elsevier's guidelines (www.elsevier.com) of what they believe constitutes honest scholarship, papers should:

1. Be the author's own work and not previously published elsewhere.
2. Reflect the author's own research and scholarship. The paper should be an honest representation of the authors' research.
3. Credit all meaningful contributions of people who participated in the study and paper including co-authors and collaborators.

4. Not be simultaneously submitted to more than one journal.
5. Cite all appropriate existing research, which places the paper in its proper context.

Judge yourself

✓ How do you feel about lying? About liars?
✓ Have you cheated in anything? What was the reason? What was the effect?
✓ Do you trust cheaters? Do you want to work with cheaters?
✓ What is an acceptable amount of fabrication in science?
✓ How do you think you'd feel if you built a research project upon others' work that you later found out was fabricated?

The case of Jan Hendrik Schön, "Plastic Fantastic"

Scientists share a common plight with all humanity: the temptation to cheat. In 2009, in the midst of writing the first edition of this book, the floodgates of news stories were open about the sexual infidelity of lawmakers across the United States. Among the lesser-known cases was Tennessee state senator Paul Stanley, who resigned from his office after reports were made about his affair with a 22-year-old intern. He admitted that it occurred. The reason I mention it here is to showcase the commentary on the affair from a Tennessee state representative Richard Floyd. He said, "You just cannot place yourself in the position to be tempted. Opportunity is the problem, and you get to thinking that the rules don't apply to you." (Associated Press 2009). But of course, we know that the rules always apply to you, me, and everyone else too.

By all accounts, Jan Hendrik Schön was a hard-working wunderkind, taking a PhD in 1997 in physics from the University of Konstanz in Germany and then moving straightaway to Lucent Technologies' Bell Labs in New Jersey. There at Bell Labs he formed numerous collaborations and worked in the area of experimental physics and materials. The breakthrough research that was apparently realized was in transforming non-conducting organic-based molecules into semiconductors. He published an amazing string of papers on the subject, which included numerous co-authors during a few years time – at an amazing rate of a major paper every two weeks! His publications in *Science* and *Nature* were heralded as significant breakthroughs that could revolutionize the electronics industry. Schön had become the rock star of science, winning many accolades and major scientific awards for his achievements until the sudden collapse in 2002. Leading up to this collapse were allegations that the results he published were anomalous and not replicable. Closer examination of the papers showed that Schön used the same plots repeatedly in different papers. When asked for his data in lab notebooks, Schön could not produce them. He claimed that because of computer problems, the data were not to be found there either. An independent investigative panel chaired in 2002 found him guilty of 16 instances of research misconduct and he was fired. Indeed, Schön had used the same set of data on several major peer-reviewed journal articles and indeed much of the data were simply fabricated. Some graphs were made using mathematical equations instead of real data. His co-authors were exonerated from any wrongdoing. In 2004 Schön's PhD was revoked by the University of Konstanz, which was upheld by a court ruling in 2011 (Vogel 2011). At that time and now PhD-less, he was working as a process engineer in German company.

The case of Woo-Suk Hwang: dog cloner, data fabricator

Woo-Suk Hwang (sometimes written Hwang Woo-suk) took a doctorate of veterinary medicine from Seoul National University and then later rejoined the same institution in South Korea as a professor. His area of expertise was reproductive biology and became known in the 1990s and 2000s for his high-profile work in cloning human cells and mammals, namely cattle and the dog "Snuppy." However, the work that received the most attention, and for good cause – it seemed as if it could revolutionize medicine – was his research on stem cells. Landmark papers were published in *Science* in 2004 and 2005 announcing the first cloned embryonic human stem cells. The first paper indicated that his group had cloned a human embryo and produced stem cells from it after 242 failed attempts (Cyranosaki 2006). The second paper, in 2005, claimed that 11 independent stem cell lines were made from just 185 donated eggs – a much higher efficiency. More importantly, the stem cell lines were supposedly genetically identical, i.e., matched, to different patients. The implications of the research to individualized medicine were immense. It was now feasible to imagine the day where people could have stem cell lines that were custom-made from each person to heal all sorts of diseases using one's own body. In June 2005, when the *Science* paper was published, there was tremendous excitement and extreme Korean national pride in the achievement.

Things went rapidly downhill (Anonymous 2005). In 2004, questions were raised about how and where Hwang acquired so many human eggs. In 2005, an anonymous source contacted a Korean television network with a tip that many of the eggs came from laboratory personnel who were coerced into "donating" for the cause. This would have meant unnecessary surgery for the women donors and would be a violation of bioethical standards. The television show, which aired in 2005, had a polarizing effect. On one hand it mobilized Hwang's supporters and buoyed national pride; Hwang had a plethora of fans. Some women even pledged egg donations. On the other hand, it caused many people to scrutinize this particular research, and research in general, even more closely. No one had replicated the 2004 work, and nor had even had the opportunity to replicate the 2005 research. Another shoe fell when a co-author of the 2005 paper, a US professor, attempted to withdraw as an author because he believed that there was significant research misconduct in the Korean lab. *Science*, however, would not allow his resignation as an author, since he had tacitly agreed to a statement that all authors in the journal acknowledge: essentially that all authors take responsibility for the integrity of the research. "Too late," said *Science*. In December 2005, rumors of data fabrication led to an internal investigation by Seoul National University, and to make a long story short, patients and their supposed cloned stem cells lacked a DNA match, indicating no cloning had happened. Further analysis uncovered image fraud in the 2005 paper in which parts of the same image of two cell lines are used more than once to represent the other nine cell lines.

The month of December 2005 was a very busy one for Hwang, his lab, collaborators, and his university. The inquiry was open and closed, with the damning finding leading to the retractions of both 2004 and 2005 *Science* papers the following month. Hwang was treated in the hospital for stress-related ailments. I doubt he had a happy Christmas. Finally, December 29, 2005, Hwang attempted to resign from his position at Seoul National University. The university did not accept his resignation and instead suspended, then fired him in 2006. Hwang was finally tried and convicted for fraud in a criminal court in 2009 and was assessed a two-year suspended prison sentence. As of 2020, Hwang was working as an animal cloner at Sooam Biotech in Korea.

There has been much speculation about why and how this misconduct happened. Could it be that Hwang had put too much of himself into his research? After all, he was a self-avowed workaholic, arriving at the university at 6:00 a.m. and not departing until midnight on most days. Could it have been that his self- and government-imposed expectations led to data fabrication? His animal cloning work has been shown to be genuine. Could it have been that he over-extended his research into an area that demanded very rapid results that were not feasible to produce? Was he simply a victim of his own unrealistic expectations? After all, there were millions and millions of dollars riding on the outcome of the research. I suspect that all of these factors played a role in Hwang's downfall and serve as warnings to all of us in research.

The case of Diederik Stapel, psychological serial fabricator

Diederik Stapel, from the Netherlands, took a PhD in psychology from the University of Amsterdam in 1997, joined the faculty at Tilburg University, where he eventually became a dean in 2010. He had various interests, such as theatrical acting, but it appears his most serious interest was coming up with ideas about things that perhaps he thought were true. One of these preconceived notions included "Coping with chaos: How disordered contexts promote stereotyping and discrimination," which was the title of a paper published in Science and was subsequently retracted. Indeed, to date, 58 of his papers have been retracted, which has been documented by Retraction Watch (retractionwatch.com). The walls came tumbling down in 2011 and 2012 when investigations concluded that Stapel had essentially fabricated data in all these papers over many years. In his own words [found in the foreword of the English translation of his book about his fabrications (Stapel 2016 English translation by NJL Brown)]:

> I've spun off, lost my way, crashed and burned; whatever you want to call it. It's not much fun. I was doing fine, but then I became impatient, overambitious, reckless. I wanted to go

faster and better and higher and smarter, all the time. I thought it would help if I just took this one tiny little shortcut, but then I found myself more and more often in completely the wrong lane, and in the end I wasn't even on the road at all. I left the road where I should have gone straight on, and made my own, spectacular, destructive, fatal accident.

Of course, he also said he was sorry that he hurt so many people, including family, co-workers, and friends as he told his story. After losing his job at the University of Tilburg Stapel published his book, did TED talks, has tried his hand at other pursuits. According to his LinkedIn page (via Google Translate), his is a psychologist and "Diederik works as a strategic advisor, concept developer, coach, co-thinker, counter-thinker, editor, mind writer" (accessed August 17, 2022).

Judge yourself

✓ What are your feelings about the Schön and Hwang cases? What about the Stapel case?
✓ How do you feel about these three scientists as people?
✓ Do you view any of the pressures they felt as legitimate justification for data fabrication?
✓ Could you envision yourself being tempted by any of the circumstances or personality traits common with these researchers?
✓ Did they receive appropriate sanctions?

Retraction Watch

Retraction Watch (retractionwatch.com) is a blog by Ivan Oransky and Adam Marcus that was started in 2010 to literally track retractions. The Center for Research Integrity is the parent organization for Retraction Watch that curates a searchable retractions database, among other things. I've found that the blogs and lists from the Retraction Watch quite useful. One way the bloggers organize stories on scientists whose papers have been retracted is their leaderboard (https://retractionwatch. com/the-retraction-watch-leaderboard): the top 30 authors with the most retractions indexed in the Web of Science, which is reproduced below from September, 2022 (Table 6.1).

Table 6.1 Retraction Watch Leaderboard as of September 2022

Rank	Retractions	Name
1	183	Yoshitaka Fujii
2	163	Joachim Boldt
3	119	Hironobu Ueshima
4	109	Yoshihiro Sato
5	88	Ali Nazari
6	84	Jun Iwamoto
7	58	Diederik Stapel
8	56	Yuhji Saitoh
9	48	Adrian Maxim
10	43	Chen-Yuan (Peter) Chen
11	41	Fazlul Sarkar
12	41	Shahaboddin Shamshirband
13	41	Hua Zhong
14	40	Shigeak Kato
15	37	James Hunton
16	35	Hyung-In Moon
17	34	Antonio Orlandi
18	33	Dimitris Liakopoulos
19	33	Amelec Violoria aka Jesus Silva
20	32	Jose L. Calvo-Guirado
21	32	Jan Hendrik Schön
22	31	Naoki Morio
23	30	Soon-Gi Shin
24	29	Bharat Aggarwal
25	29	Victor Grech
26	29	Tao Liu
27	28	Cheng-Wu Chen
28	27	A. Salar Elahi
29	27	Prashant K. Sharma
30	26	Richard L. E. Barnett

Detection of image and data misrepresentation

Falsified or fabricated data and images can be detected during stages of publication and grant proposal review. For example, grant proposal and manuscript reviewers might see potential issues in submitted documents and report them to grant and editorial authorities, who then follow up. Before we explore the mechanisms and tools for detection, let's backtrack to formally define fraud. Fraud is the purposeful attempt to deceive. Most scientists seem to be uncomfortable with the term "fraud" in that it does imply intent. Fabrication, falsification, and plagiarism (FFP) are the categories of activities that the US government's Office of Research Integrity defines as misconduct. They are more concerned with whether or not a scientist they funded did these deeds than ethics per se (Shamoo and Resnik 2003). If data are made up or manipulated in such a way that is designed to draw interpretations or conclusions that are not true, then research misconduct has occurred. There are many publications that go into great depth detailing nuances of misconduct. Two publications of note are a 2009 National Academy of Sciences report entitled "Ensuring the Integrity, Accessibility, and Stewardship of Research Data in the Digital Age" and "What's in a picture? The temptation of image manipulation," an article written by two editors of cell science journals (Rossner and Yamada 2004). For our purpose here, my simple objective is to briefly examine what is allowable and not allowable for the digital manipulation of images, and then to understand how deviations from accepted practices might be detected.

First of all, I readily admit I'm a dinosaur when it comes to image manipulation. When I was a graduate student (late 1980s–early 1990s), we ran gels and took Polaroid photos of them. The developed picture of the gel or Southern blot was then literally pasted with glue onto a piece of paper that contained the figure labels. Then, another picture was taken, typically using a 35-mm camera, again, with film. Rolls of film were sent out for developing and the result was physical slides or prints for use in talks, papers, and grant proposals. Typically, multiple exposures were made in taking the photographs to give the best chance of having esthetically-pleasing figures. At the time, all this photography for image capture seemed to me to be an art that I wished to avoid, hence the big numbers for repetition, given that

I was far from an artist. Laboratories with good funding would outsource some of the art to studios within the university. As a result, these professionally -rendered pictures were discernibly prettier than my do-it-yourself images. Images of microscope slides or organisms were treated similarly. Sometimes pictures were drawn freehand and a photo was taken. What you saw was what you had. I suppose there would have been a way to doctor them to show something different, but it never occurred to me to do so. Essentially, all the images and the entire manuscript or grant proposal was submitted on paper and not electronically as they all are today. Multiple paper photocopies in fact, with glossy figures were submitted to journals or granting agencies where the copies were subsequently snail-mailed out to reviewers.

I do recall wishing that I could make my data-based images look better. I felt like the visually artistic scientist had an advantage over me, the artistically-challenged scientist. The background hybridization on a Southern blot and the ill-contrasted microscope picture from my own hands were sometimes downright ugly. When published, it might have sent the message that my science was a bit sloppier than ideal or that I didn't care about appearance as much as I should. The only real alternatives to submitting the not-so-great pictures were to repeat an experiment and hope for cleaner results and better pictures or to not include certain data and hope that the story would be sufficiently told. On more than one occasion a reviewer pointed out that perhaps an experiment needed to be repeated to obtain a figure that would be of "publication quality." I hated reading that since it meant repeating an experiment that to me would only result in a prettier picture and not better data. I recall the drive to move on from art to performing new experiments to generate new knowledge, and not simply more attractive images.

As I think about it, I hardly see any ugly data pictures in journals these days. Does that mean the data are all cleaner and experiments better, or is something else going on? Today, the temptation to manipulate images using Photoshop or other image software must be much greater than in the good old days of physical pictures on paper. A researcher can now "clean-up," crop, merge, highlight, label, and outright change images to suit any whimsy. What is allowable and disallowable? What are the best practices?

The overriding principle for images is that no action should be done to an image that could lead the reader to an interpretation or conclusion that the raw picture or data do not warrant. Obviously, features should not be added or deleted from images. If brightness or contrast changes must be used, they should be used cautiously, and then to an entire image. Non-linear correction, such as gamma correction, should probably never be used, and if done, it should be disclosed in the figure legend (Rossner and Yamada 2004). Essentially all image manipulation should be disclosed in the text of manuscripts. Yet, I don't see much disclosure in publications.

How can figure fraud be detected? It might be difficult for the typical reader or reviewer to see problems in manipulated figures, but editors and office staff who handle a lot of manuscripts acquire an eye for data and figure manipulation. Some people in my lab have found duplicated images from previously published work in submitted manuscripts without even looking for them. In one instance we were really interested in a certain new area of research, and so we jumped at the chance to perform a peer review of a tissue culture paper using one of our new favorite plant species. In this submitted paper, the authors cited one of their previous papers. In the previous paper, they stated that they'd use one set of constituents in their tissue culture media. In the new paper, they used another set. We were happy to have discovered their previous paper, since we weren't aware of it before we performed the peer review of their new paper – they cited the old paper. Lo and behold, we noticed that there were duplicate images common in the two papers. Publishing the same images in two different papers is also not allowed, because data and displays are assumed to be novel and unpublished at the time they are submitted. They are assumed to represent different experiments. After all, if they are identical, why publish the second paper? This requirement of novelty in publishing is especially true of peer-reviewed research papers. It is also often a copyright violation to reprint pictures without the permission from the copyright holder, typically the journal or authors. In this case, however, the authors were inferring that the tissues pictured in the first figure were those described in the second paper. Of course, the same pictures appearing in the first paper claimed the same thing about the first study. Since the pictures were identical, it was impossible for at least one of their claims to be true. So, as peer reviewers we alerted the journal editor of the misconduct.

Once contacted, the authors might have replied to the editor, "This was just a mistake. We intended on using another figure and inadvertently included one that was already published. We have the correct figure – here it is." While this explanation might indeed be factual, these authors' credibility is now quite low in our eyes. We'll naturally be on the lookout for other papers by them and, naturally, we'll scrutinize their science for FFP. As peer reviewers, we have seen this type of misconduct now twice and found it when we were not even looking for research misconduct. We were simply gathering information for our own science and happenstance led us to the duplicate figures.

Indeed, for the WooSuk Hwang case, the editor of *Science* did not believe that it was likely that reviewers could have seen the problems in the submitted manuscript (Couzin 2006b). "Peer review cannot detect [fraud] if it is artfully done." The paper was examined by nine reviewers – six or seven more than is typical for most journals. Plus, the editors read and reviewed the paper. Editors and reviewers may or may not be on the lookout for misconduct. One thing that may cause someone to look for data irregularities are other faulty items in the manuscript or paper. In many ways, scientists are a trusting lot. They generally expect that honesty is a shared value among fellow scientists. Editors should probably not be so trusting, especially when image misrepresentation is so easily carried out.

Even though image fraud can be sometimes be detected by eye, wouldn't it be helpful if some dedicated people and dedicated software could help detect FFP in more systematic fashion? We have seen iThenticate in action. It is a fabulous tool for plagiarism. Is there an "iThenticate for image fraud?" Indeed, some editorial offices are examining figures more carefully and subjecting them to image analysis for the purpose of detecting manipulated data. Image analysis can detect if a composite figure has been produced, whether elements have been added or deleted, and whether a figure is "too good to be true." One such set of tools is the Office of Research Integrity's Advanced Forensic Image Tools for Photoshop (https://ori.hhs.gov/advanced-forensic-actions). These can be used to detect irregularities within images as well as comparisons between two images. The latter feature would help identify duplicated images or fabricated images. Cardenuto and Rocha (2022) identify several forensic automated analysis tools and benchmarked their performance. Essentially, they found that there is plenty room for improvement needed for robust automated screening for image manipulation. They also noted that artificial intelligence (AI) approaches for creating "deep fakes" (Gu et al. 2022) may outrun the ability to detect, but AI approaches may be useful in forensic detection, say for finding duplicated images (Van Noorden 2022). Clearly, this is an area of research to monitor in the future.

Perhaps one of the best "tools" for detection is a human with an adept eye to detect problems. Sometimes I joke that finding image manipulation is my "superpower," but my capabilities pale to those who pursue such quests in more earnest and time than me. One such person is Elisabeth Bik, a one-time biomedical researcher turned full-time research integrity investigator. She is among several scientists who have taken a keen interest in ferreting out misconduct, especially image manipulation and duplication. Dr. Bik has found problems in hundreds of papers over the past few years (Shen 2020). Indeed, she mainly uses her visual talents to find problems with published papers and then uses various venues to call out the issues she's discovered, including Twitter and PubPeer. Indeed, PubPeer.com ("the online journal club") has become an outlet for posters using their real names, pseudonyms, or anonymously, to post perceived problems in published papers. It has become the de facto virtual whistleblowing platform in scientific publishing. To wit, Dr. Bik has attracted some negative comments from her activities as well as legal threats (O'Grady 2021). On the other hand, she has tremendous support among the scientific community. When Dr. Bik speaks, people listen.

With data integrity at stake, it is important that researchers are honest with data. As Rossner and Yamada (2004) point out "creating a result is worse than making weak data look better," ... but these are both forms of misconduct. Accurate representations of study are clearly expected by readers and the science community.

Judge yourself

✓ Would you be tempted to overly manipulate an image to make it prettier or to avoid repeating an experiment?
✓ How important to you is the esthetics of scientific images?

Lessons learnt

Why do up and coming science superstars like Jan Hendrik Schön and WooSuk Hwang cheat? And cheat at the highest levels? By all accounts both of these researchers were competent and productive prior to cheating. Did they believe that the means justify the ends? That is, perhaps what they believe to be truth really is true, and the pressure to establish the facts using science is too slow for their satisfaction. Perhaps they believe the rules of science apply to "ordinary scientists" and not themselves. Perhaps they became addicted to publishing in *Science* and *Nature*. Whatever the case may be, and I think there are multiple causes, research misconduct is infuriating to both the straight-players in science and the public. In all of the cases we examined, sloppy or fraudulent science cost innocent researchers much time and effort in trying to replicate the unreplicatable. These innocents are typically graduate students and postdocs whose time was wasted and careers delayed. And what about innocent parties within the misbehaving PI's lab? They are now often deemed guilty by association. After all, who would even trust a letter of recommendation from WooSuk Hwang? The price of research misconduct is hefty, and its turbulence runs deep with collateral damage.

So, how are we supposed to respond if we detect research misconduct? I gave one brief example in the case of reviewing submitted papers for journals – simply alerting the editor of a probable misconduct. We'll look into both the more common forms and few less simple (and more painful) examples of "whistleblowing" in the next chapter.

Summary

- ✓ Fabrication is probably the most heinous of research misconduct and can cause damage to the perpetrator's career, but also to those of trainees and other scientists who trust research findings are true.
- ✓ There are judgment calls about which data to include in a report, but data fabrication data is never allowed.
- ✓ Be careful when manipulating figures to avoid misconduct. Declare all manipulations. The data shown need to truthfully represent the data collected.

Chapter 7

Research Misconduct: Falsification and Whistleblowing

ABOUT THIS CHAPTER

- Falsification is making false statements on pertinent documents, e.g., grant proposals and papers, and is considered to be research misconduct.
- One of the most difficult decisions a scientist makes is when and how to report a case of research misconduct.
- The decision has important consequences for the reporter and reportee as well as other people in the lab and university.
- Reporting misconduct discovered as a reviewer is easier than blowing the whistle on a colleague, but this should not be done casually.
- Science integrity is our corporate responsibility, but there are many procedural considerations to be weighted when contemplating blowing the whistle.

Integrity is tested when misdeeds are observed and a scientist is left with a difficult decision: report misconduct, or not? How is it done? What about the special instance in which a student's colleague or mentor is making up data? What are the consequences? Whistleblowing is not for the faint of heart or the coward, since there are many documented examples of retribution and unintended negative consequences that do not favor the whistleblower. Indeed, great courage is required by responsible scientists to face unpleasant facts and do the right thing for the sake of science integrity. This chapter is perhaps the most important one in this book inasmuch as it, in many ways,

---

Research Ethics for Scientists: A Companion for Students, Second Edition. C. Neal Stewart, Jr.
© 2023 John Wiley & Sons Ltd. Published 2023 by John Wiley & Sons Ltd.
Companion website: www.wiley.com/go/stewart/researchethics2

defines professional ethics in research science. The chapter integrates mentorship, responsible conduct in research, grantsmanship, research pressure, with the responsibility we all share as citizens in science.

Whistleblowing takes its meaning from reporting wrongdoing, e.g., a crime in society or breaking the rules in a game, say, when a bobby or police blows a whistle on the criminal or a referee whistles a foul in sports. In all cases, the perpetrator is not very happy with the whistle being blown and many other people are often dismayed as well (e.g., think about your reaction when your favorite football team is called for a penalty). Almost nobody seemingly appreciates the whistle-blower. If the person being reported is popular, personable, and had apparently lived an exemplary life in the past, there is a tendency to judge that the whistleblower is simply wrong or vengeful. It is true that being wrongly accused of wrongdoing is hurtful toward one's reputation and career. Whistleblowing is not something to be done without great consideration. The possibility of errant whistleblowing is also threatening and a factor in non-collegiality. After all, if the whistle can be blown on my colleague, then perhaps I'm the next target. Reporting FFP should always be done after very careful contemplation and with due diligence.

We will examine one of the most publicized true cases of whistleblowing in recent memory and its ramifications on the people involved. In this case a University of Wisconsin professor was reported for research misconduct by her graduate students. The case is true and documented in at least two reports (Couzin 2006c; Allen 2008). It does not have a happy ending for any of the involved parties. To a person, their studies and research projects were severely altered or abandoned, and in most cases, they left their university studies under duress.

Reporting and adjudicating research misconduct

Understanding the typical structure for maintaining research integrity is important when thinking about blowing the whistle. In universities, the chief research officer is ultimately responsible for investigating accusations of research misconduct. In companies and research institutes, a parallel position would exist for that duty. To that end, the institutional research integrity officer (RIO), who ultimately reports to the chief research officer, is on the point to investigate and adjudicate accusations of research misconduct. Typically, universities do not have standing committees or panels to investigate and judge cases of alleged faculty misconduct, which contrasts the situation with students. Most universities do have undergraduate and graduate honor systems. Why the difference? It is simply a matter of frequency of cases. It is rarer that cases against faculty crop up. We will see why later. There are many routes how a potential case gets to the RIO. Most often research funding is involved, and anyone who works on funded research is held accountable by the agencies via the university. Those held responsible include administrators, faculty members, postdocs, graduate students, and undergraduate students. If the RIO determines there is possible merit to an accusation, i.e., that it should be investigated, the people who may be responsible for committing misconduct are identified. They are known as the respondents. The person who accuses is known as the complainant. The particular work that is investigated may be a grant proposal, a published article, or an unpublished manuscript that authors may intend on publishing, or some other scholarly work. In the end, respondents could be found "innocent" (not responsible), or "guilty" (responsible) of misconduct. Essentially for it to be deemed as misconduct, the respondent would be found reckless, knowing, or intentional in the charge(s) of misconduct where reckless is the least bad and intentional is the worst. Honest error would never be considered to be research misconduct.

Before we get to this most painful case of internal whistleblowing, we will examine external whistleblowing, which may not be as angst-rendering for the institution and units, should still be done only after facts are clear and wrongdoing is reasonably certain. Scientists are responsible for maintaining truthful reporting of data; their own and others. That is our corporate duty as scientists. External whistleblowing is much more common than internal whistleblowing, i.e., reporting a professor within a shared institution is rare. As an external whistleblower, I have notified editors and grant program managers of likely problems several times, but have only been an internal whistleblower once. There are two overriding reasons for this. First of all, I have more opportunity to see problems with papers or grant proposals in my roles in the larger universe of science than misconduct perpetrated by my own colleagues. Second, the author of a bad paper or grant proposal outside my institution might not be known by me. A colleague is more likely to be a friend, lab-mate, or both, and so the dynamics are different. Another reason has to do with the nature of the editor/grant officer-reviewer relationship. The reviewer is basically charged with finding errors in submitted papers and grant proposals. That is, reviewers are invited to dig around to unearth issues. While most reviewers don't specifically look for FFP (I typically don't), sometimes they find it anyway and, then, should report it along with the normal errors that all reviewers denote while reviewing. As an editor, I do look for manuscript irregularities, including potential FFP. As we'll see in a later chapter, all peer-reviewing done for papers and grant proposals is typically performed anonymously; so ethics would not even allow me to informally intervene in a paper submission, even if I wanted to. For external whistleblowing – a paper or a grant, the whistleblower is nearly always completely protected from any negative ramifications of reporting misconduct; i.e., whistleblower fallout and retaliation. The editor or grant officer may or may not constructively do anything with the reported information, but at least the whistleblower is not penalized. The structure of reviewing that features anonymity does favor objectivity and honesty while protecting the reviewer. Internal whistleblowing in not so straightforward; it is definitely not painless.

A "can of worms" indeed: the case of Elizabeth "Betsy" Goodwin

The profile of a successful professor

> "Goodwin was one of the rising stars of the UW's genetics department. 'She was an extremely good citizen,' says Phil Anderson, another *C. elegans* expert and a professor of genetics [at the University of Wisconsin-Madison]. 'She did more than her fair share of committee work, and she was very involved socially. She entertained prospective graduate students and helped recruit new faculty. She brought a genuine sense of joy to working here'" (Allen 2008).

Dr. Elizabeth Goodwin had received her PhD from Brandeis University and completed a postdoctoral stint at the University of Wisconsin. She had published over 50 papers, some of which appeared in very high-profile journals, had been successful in grant-getting, and had a vibrant lab of six PhD students and a technician. Using every available indicator of success, Goodwin passed with flying colors. In addition, her group also had fun together outside of the laboratory walls. They enjoyed parties and horseback riding adventures among other normal extra-science functions that bring lab "families" together. Professor Goodwin was a hands-on mentor, communicative, and called regular lab meetings. The scientists were united in working on a suite of challenging problems at the cutting edge of science.

Her lab was one of the few looking at sex-determination in the model nematode "the worm" *C. elegans*. Every person in her lab actively studied this organism, which is a very powerful developmental biology model. As it turned out, Goodwin hypothesized that one particular small RNA gene was crucial for sex determination. The gene had been linked to the male/female trait in *C. elegans*, but the small RNA hypothesis was quite novel. As we'll see, her dogged determination to follow this line of research might have led to some of her ensuing problems. Small RNAs were (and still are) a hot topic since they control gene regulation in some instances. Another pertinent fact is that five of her six PhD students had been working in her lab a number of years, but with little data to show for it. So in many ways, this case is a "perfect storm" of dynamic researcher, a hot research topic, the need for continued funding, but with little recent success.

Goodwin gone bad

Our story begins in the autumn of 2005 (Couzin 2006c). Chantal Ly was a frustrated student in the seventh year of her PhD program. She was unable to replicate certain experimental results gathered in the Goodwin lab. I've been there with my own advanced students. You can hear their clock ticking – the degree that should have taken far less time to complete is stymied because experiments are just not working out the way envisaged. It is never an easy situation. Dr. Goodwin then does something that I've had success with my own trainees – she gives Ly part of a grant proposal to spur an interest in pursuing something else that might work, which could then substitute for the unfruitful research. We all want students to experience success. Goodwin posits the proposed research as experiments she'd intended for another PhD student, Garett Padilla. Goodwin thought, however, that there would be room for both students on the grant, should NIH choose to fund it. After the proposal was submitted Ly noticed that there is a figure in Goodwin's proposal labeled as unpublished data that was, in reality, published by Goodwin and co-authors the previous year. That part is ok. But the protein of interest in the published paper was different from the one indicated in the proposal. That part is not ok. Ly was so troubled that she shared the portion of the proposal with Padilla, who was also her officemate, to ask his opinion about the possible falsification. As they looked further, they noticed more irregularities in the grant proposal. It had appeared that Dr. Goodwin had falsified data in more than one instance. What would be the student's next step? One certainty was that disillusionment rapidly set in as the students learned that all was not perfect in paradise.

The grant treadmill

Grants are hard to win. Typically for any funding panel, as few as 10% of proposals submitted result in awarded grants. Since it is a competition, professors must "sell" their proposals to boost their chances of winning. There is a temptation to approach the edge of truth, and sometimes go beyond it. This practice breeds skepticism in reviewers who must cut through any potential smoke and mirrors. Proposal reviewers want to see preliminary data that convince them that an idea and approach are sound and merit funding. In the Goodwin case, it appears that the preliminary data went over the line of acceptability and into the world of misconduct. But why would Dr. Goodwin feel as if she needed to cross the line? In two words: competition and doubt. As scientists, we all know that our proposals are competing against other proposals, therefore there is the potential temptation to push the envelope. It is an unfortunate part of the system of science, but no one has thought of a sustainable alternative to competition to support large amounts of the best research. The assumption is that the best science (best proposals) typically receive the funding. Doubting your proposal's ability to compete is an ever present demon since you don't know anything at all about the competing proposals. You wonder, who's submitting what? Who's on their team? How will the reviewers compare my proposal with the competition? This dynamic situation creates pressure for researchers. Proposers must diligently stay in the right state-of-mind to stay on track. In a research university, pay for graduate students, postdocs, and other personnel, supplies, equipment, etc., may be supported solely from grants. There is definite motivation to write winning proposals. The alternative is a starving lab; no mentors want their researchers to starve. In some areas of science, the pressure for funding is greater than others; the more expensive the science, the more pressure for funding. Small amounts of funding needed to grow and weigh plants is not as stressful as that needed for sustained biomedical research.

Judge yourself

✓ How do you handle pressure and competition? Can you stay cool or does it keep you awake at night?
✓ Do you enjoy "selling" and being the person in charge?
✓ Can you trust in the "system" and play by the rules?

The plot thickens

Padilla started keeping a log of the case (spoiler alert: he later goes to law school) and then consulted with another UW faculty member, a person in another department with whom there were personal connections. He also consulted with a former Goodwin postdoc who was, by then, working at a company. They both encouraged Padilla to talk with Goodwin about the matter, which he did – on Halloween [cue the scary music]. According to Padilla (Couzin 2006c), the meeting did not go very well. She denied falsifying data, but admitted that mistakes were made. At the end of the meeting Padilla did not feel as if any satisfying resolution had ensued.

After this meeting, all seven lab members were invited to a meeting to discuss the alleged falsification that was held in a building different than that housing their lab for discretion to discuss their plan of action. Choice 1, having a discussion with the UW administration, was dismissed as being too risky. They wondered what would the administration do to Dr. Goodwin? How would it all be handled? Choice 2, sending Padilla back to speak with Goodwin, seemed like a more prudent path. After all, perhaps Padilla, the lawyer-to-be, could discover the real underlying problem and solve it. Maybe she could retract the grant proposal. Indeed, the second Padilla–Goodwin meeting seemed to go better. She promised to email her NIH contact and copy Padilla (which she did) explaining the problems in the proposal. She asked for Padilla's forgiveness. She said there would be little chance the proposal would be funded. "No harm, no foul," thought Padilla.

Judge yourself

✓ How would you feel to by in a situation similar to Ly and Padilla?
✓ Are you, by nature, confrontational, or do you avoid confrontation and just seek to get along with everyone?

Not so fast

Couzin (2006c) tells of a third student, Mary Allen, who was not convinced the situation was resolved sufficiently for her to conscientiously allow her to remain in the lab. Even though she was in her fourth year of a PhD program, she was willing to change labs to remove herself from the problematic situation and mentor. After explaining her wishes to Dr. Goodwin, additional reasons and explanations were forthcoming from Goodwin as to what had occurred. The professor claimed to have received unlabeled images from a lab mate and had simply mixed up the files. Goodwin admitted to making mistakes, perhaps suboptimal judgments, but she was adamant that she had not falsified data. After this meeting, the students continued to confab and as a result, grew more and more worried about the future. The proposal to report Goodwin to the administration was forwarded again. Instead of dismissing this idea outright, they decided that they would not visit with administrators unless it was unanimously decided by the lab to proceed. They were also careful not to discuss the matter with others outside the group, except the two people they'd already brought into their inner circle as advisors. As November turned to December, they finally made the decision to report Dr. Goodwin to their department head. The students reasoned that, even though it might go bad for them, Dr. Goodwin might place future graduate students in their same predicament if they did nothing. We can see that by their actions they were obviously convinced that she was guilty of research misconduct. Their decision was a brave one, and – important – not a compulsive one.

Not a great Spring semester for the professor

The department head referred the matter to two deans, which informally investigated to see if there was just cause for further investigation, which is a reasonable path for the administration, even though students, department chair, or dean could have simply referred the case to the RIO directly. In a meeting that included the department head, Goodwin, and lab members, Goodwin claimed that she "was juggling too many commitments at once." This, certainly, is a common predicament for faculty members to find themselves. A faculty member has tremendous pressure to find and keep funding, publish papers, mentor students and be a good citizen of the university and to science. She claimed the bad figures were just placeholders for good figures she had intended to substitute in the proposal. I wonder about how the figure legends could be explained? It seemed that the explanations simply did not jibe with reality. The informal investigation turned into a formal investigation with the RIO appointing three professors charged to uncover the truth. Goodwin grant proposals that were funded in the past were also inspected. It turned out that funds from all three grants were sent back to the NIH and USDA.

In the end of it all, Elizabeth Goodwin resigned March 1, 2006, not even halfway through the Spring semester. UW officials were obliged to continue the investigation to assure that they had uncovered all the problems so they could report the findings back to the granting agencies. This task was made more difficult by Dr. Goodwin's absence. Another notable fact is that the US Federal Bureau of Investigation (FBI) became involved (Winter 2010).

The fallout

Even though their professor was largely absent all semester, even before her resignation, the students attempted to carry out their research and lab-life as usual. They did experiments and held lab meetings sans Goodwin. They were assured by the administration that their salaries and stipends would not disappear that semester. And indeed, all of their funding remained intact during the short term. But a lab without a PI is like a symphony orchestra without a maestro – Tchaikovsky's Fifth Symphony is not an option without the maestro and neither is high-level science possible without the PI. Except that this musical analogy underestimates the situation. Graduate students cannot complete their degrees without a mentor. Whether the mentor dies, resigns, or does not get tenure, there are limited choices for graduate students to complete their research projects and degrees. When I switched universities during the midstream of two graduate students' degree programs, they both moved with me. I'm not sure what their options would have been if they'd decided to stay at my first university. If a professor simply resigns or dies, students can either find another professor to finish them up or they can start over. These choices were the ones that awaited the three graduate students, but there also seemed to have been an element of "contamination." University departments are cozy entities. Few people want to risk contamination of "damaged goods" into their labs My choice of words here are very stark, but in the worst of situations, many faculty members would have this perception. In the best of outlooks, the system of graduate student education at most universities is simply not set up to handle deviations from the traditional mentor-student relationship. Even there, the relationship can become strained. Few faculty aspire to enter into a mentor-student relationship that is odd from the outset.

My second-best favorite quote from the Couzin (2006c) article was about how many of Goodwin's faculty colleagues felt about the situation. "Goodwin had had 'to fake something because her students couldn't produce enough data.'" I'm sure the faculty rumor mill had been working overtime in attempt to explain the problem away and exonerate Dr. Goodwin. Faculty members are famous (or infamous??) for looking out for their own. Human nature demands that they likely didn't want to face the fact that their colleague was trying to prove something that was not scientifically possible and that fact was the real reason for the students' unproductivity. So, given that the above view was rampant, I don't imagine the UW faculty were falling over each other trying to recruit the students. Even if they were, the students would mostly have had to start over with new projects. However, two of the students did stay at UW and had successful "do-overs." The other four PhD students left. Of these, only Mary Allen continued on as a PhD student. Padilla went to law school and Ly moved to working at a company.

Judge yourself

✓ Did the Goodwin debacle end the way you think it should?
✓ What are the alternative endings to the story?
✓ What would you have done differently if you were in the students' position?

Life is so unfair ... to the whistleblower. Can you blow the whistle and still be a respected scientist?

In retrospect, the students mostly did everything "right" with regards to ethics but enjoyed few, if any, direct benefits from being good ethical citizens in science. They did not take lightly their difficult decision to blow the whistle. They kept, to a large degree, the situation confidential. Finally, after the students decided to report, they reported to an appropriate university administrator; their department chair. For their efforts, their degrees were largely scuttled and the time and effort they had devoted in pursuit of the PhD was wasted. I'm not sure how it could have ended differently, however. Such is the life, perhaps, of the student whistleblower. But they all fared better than Goodwin. She was fined $500, ordered to pay a total of $100,000 restitution, and was sentenced to two years probation after being found guilty of making a false statement (Winter 2010).

Science is not alone with regards to unfair treatment of whistleblowers; police departments, schools, politics, and businesses have all seen unfair persecution of people who point out injustices and wrongdoing in the professional world. Retaliation can be brutal. As Gunsalus (1998) points out, society has a "visceral cultural dislike for tattletales." Indeed, if the whistleblower gets it wrong or is vengeful, an innocent person obtains a sullied reputation or worse. Thus, the motivations and conscience of the teller are important; from the outside, these might not be evident or even knowable. Certainly, believability, and trust in the whistleblower are key, regardless of motivations. This is why the utmost care is needed when assessing such a situation and taking action.

Take, a fictitious example of two evil brothers. When one evil brother tells their father about the evil deeds of the other brother, the dad is often dismissive of the event. Yes, he might believe that the perpetrator had done an evil deed, but the whistleblower, also evil, has low credibility as a reporter. What about the situation when just one of the two brothers is evil and the other is good? When a good brother tattles on the evil brother, experience has shown that the good brother is usually trustworthy and right. In either case, however, the father might prefer for the brothers to come to justice without his involvement. Let's contrast this case of brothers to a case in science, such as the Wisconsin incident. No matter how you slice it, it is impossible to paint a scenario where graduate students can benefit from their major professor's professional demise, especially when the group of students act together to report. It matters little whether the graduate students are viewed as good or evil. Therefore, perhaps we can eliminate any motivations of retaliation or personal dislike in such a case, since it is certainly to the students' disadvantage to report their professor's misdeeds. Even if their ethical motivations prevail, they still could have been mistaken in their accusations, but again, since they acted in concert with due diligence, this scenario is unlikely either. We are faced with the facts that it is very unlikely that any graduate student or postdoc will unjustly report his or her mentor for anything unless they've left the university as a disgruntled former employee. I recall a certain graduate student whose mentor had caused him all sorts of personal pain and anguish during the course of his PhD degree program. In a particular moment of brutal honesty coincident (and after a few glasses of wine), the student confessed that he would not hesitate to murder his mentor by strangulation, except that he wouldn't get his PhD if he did it. Nonetheless, dishonest whistleblowers exist in science.

A university administrator's opinion

While universities might wish to protect whistleblowers who do the right thing and report serious cases of misconduct, the deck is stacked against the reporters, especially when the reporters are students. Administrators know that. Let's face it. A graduate student, whose role is most likened to a type of intern or apprentice, is in a position of great inherent weakness relative to his major professor, especially if the professor holds tenure. While it may seem that professors are also relatively weak compared to administrators, this is not necessarily the case since tenure is powerful in maintaining the *status quo*. One important role of society is to protect the weak. Why should a university society be any different from society at large? As we observed in the case above, the close mentor–trainee relationship of graduate education results in the intertwined destinies of professor and student, especially in the case of failure by the professor. In addition, this relationship must be built upon mutual trust. I can't see any way around this situation in today's graduate education system. Considering these circumstances and others, let's look at the Gunsalus (1998) six rules for responsible whistleblowing.

Rule 1 is to consider various explanations about a potential incident, including "especially that you [as the whistleblower] may be wrong." It is critical that a complainant think big and broadly about an incident of misconduct. It is important to understand all the facts and to be able to explain them. None of us are omniscient, and there are typically mitigating circumstances to consider.

Rule 2 is critical: "ask questions, do not make charges." No one likes to be accused of wrongdoing, and in light of Rule 1, that the reporter could be wrong, it is important to exercise the student's inquisitiveness rather than the victim's judgmental side. In the midst of what could eventually be whistleblowing, it is important to ask as many questions as possible and to make known that questions are merely being asked. You don't understand and you want to. Listening is more important than talking when asking questions.

Rule 3 is to focus on facts and data. Without these, there is no case and people are more apt to focus on the whistleblower than on the whistleblowee and the potential misconduct. If the data are not available to the reporter, they might or might not be available to others. Evidence is crucial and cannot be overlooked. Making false charges may be a career-killer to a complainant.

Rule 4 is to separate personal and professional concerns. Students seek to be professionals in science, and it is important to conduct investigations into FFP in as factual way as possible, even when emotions come into play. This is not to say that a complainant must be dispassionate since professional ethics is something that many people ardently protect. It is important to focus on deeds (or misdeeds) rather than the personality of the perpetrator or your own personality.

Rule 5 is that goals should be assessed. A potential whistleblower might wish to simply change a situation informally rather than lodging formal charges; behavior change could be more important than punitive charges. Of course, in light of the fact that misconduct should never have happened to begin with, one objective of the reporter is to ensure that it should not happen again with this mentor. Here an administrator could reasonably ask a complainant what they want the administrator to do about it. These goals should be elucidated.

Rule 6 might be the most important of all: "seek advice and listen to it." A student should realize that university administrators have seen more problems and have a larger scope of professional insight than does the student. Other people might also give advice: other students, trusted colleagues, relatives in relevant positions; but the confidence concerning the potential respondent needs to be maintained.

In summary, it is important that a potential whistleblowing student learn the facts, ask questions, seek advice, listen to the answers, and be professional.

In light of these rules, Gunsalus (1998), an administrator at a large research university, offers step-by-step procedures for responsible whistleblowers. Again, I will frame these steps with the student or postdoc in mind.

The first step relates to Rule 6: share concerns with a trusted person. Everyone needs to a trusted colleague who can lend a non-judgmental ear. This process allows the potential reporter to understand the gravity of his or her position while simultaneously gaining insight and advice. It is a rehearsal of formal complaint too. Again, asking questions rather than leveling charges is a powerful and prudent approach. It is very important that a complainant not embellish or exaggerate perceived misconduct of anyone. Again, a non-credible narrator will do more harm than good. A large part of trust is confidence, so it is important to be assured that information will remain confidential. It is also important to choose your trusted confidante very carefully. You would not want to choose someone with a conflict-of-interest or a personal friend of the respondent.

The second step is to listen to the advice of your trusted colleague. The person might agree or disagree with you and your planned action. Either way, it is important to really listen and to be objective. Listen to what you say through another's perspective – gain their ears if you can. Try to read the situation and your trusted advisor's comments. Take them seriously. If a reporter still holds a strong opinion that an official complaint should be lodged, then the next step must be taken, but carefully.

The third step is "to get a second opinion and take that seriously too." Gunsalus makes the point that even large research universities are small communities. It is easy to "get on a roll" of "confiding" in people so that a gossip chain is initiated. That is the last thing a potential whistleblower needs. My guess is that administrators who make their living from listening to complaints make each other aware of complaints, which is to say, they are networked within a university. That is not a gossip chain, but could play a role in how the whistleblower may be perceived. My guess is that such professionals can "sniff-out" a disgruntled troublemaker on a mission. Like Step 2, listen to the second person's opinion, how the facts are reflected, and focus on asking questions not making complaints at this point. This might be a teller's last real chance to tell the story without long-lasting consequences. Formal investigations are not for the faint-hearted and to be sure, there will be unintended consequences for the whistleblower.

Step 4 occurs once a decision is made to initiate formal proceedings: seek strength in numbers. As we saw with the Wisconsin case, the entire lab group together reported to their professor. If there are natural groupings of people who share your concerns, then perhaps they would also share responsibility as complainants. Again, sparking off the rumor mill should be intently guarded against. It works against the process and the whistleblower by unnecessarily harming the person to which complaints might be made. Gunsalus makes the point that if people don't wish to join in a complaint that the primary complainant should assess to answer why not. It might be informative as to how an official might receive the case and should be factored into whether to go to Step 5.

The fifth step is to find the right place and procedure to file a complaint. As a student, a department head is almost always a good starting point. The head can illuminate procedures and again, act as a sounding board. It might be that a potential complainant will be dissuaded from going to Step 6. It is important that this advice should also be seriously considered. Administrators almost always have more experience in this field than graduate students.

Step 6 is officially reporting a concern. Again, this should focus on the facts, a question, or neutral observations. It is best to avoid acting as judge and jury. As much of the evidence as available should be given. The RIO's job is to find other evidence as needed. This fact-finding is a disruptive process. Lab notebooks, computers, and facilities may be temporarily sequestered as data are copied. Everyone touched by the laboratory will be affected. Productivity will decline for everyone. It is a LOT of work to sequester evidence and cooperate with the procedure. The potential complainant needs to seriously consider these ramifications.

The seventh step is to take notes and ask questions. It is important for a whistleblower to self-protect by keeping informed and active in the process. Keep a journal of the same quality as you would a lab note-book with dates and numbered pages. Gunsalus points out that it should not be necessary to seek council of an attorney. In addition, never speak to the press! And whatever you do, avoid any mention of any complaints on social media. A blogger can go from complainant to respondent or (even worse) a defendant in a defamation lawsuit. While free-speech is a protected right in most countries, slander, libel, and defamation are not.

The last step is to exercise patience. Investigations and procedures will take some time to complete, as they should. Stay involved and be patient. Hope the FBI doesn't get involved!

I will add one additional step in addition to Gunsalus's excellent list. Any student who decides to go the route of filing a formal complaint should have a support group. After blowing the whistle life will not be easy for anyone. Being a graduate student is already hard. Becoming a graduate student who is perceived as being ungrateful, not loyal, and other unpleasantries (fill in the blank) will likely need all the personal help they can get. And they will also need to look for another mentor, likely to be located far away.

Judge yourself

✓ Do you have a support network? Are there people in your lab or network that you trust to share sensitive information?
✓ Do you know your department chair and dean?
✓ Do you have the kind of relationship with your major professor that enables frank discussions?
✓ Would you ever envisage yourself as a whistleblower? How bad would an event have to be to report it?

Deal with ethical quandaries informally if possible.

Case study: Jill and the uncloned gene

As we see from the actual case study, becoming a formal complainant typically results in a one-way path to a place most students and postdocs did not initially chart for themselves. Therefore, my best advice for potential internal whistleblowers is to proactively respond to potential misconduct issues early, quickly, and informally. This procedure also takes courage, but can reflect much better on the informal than on the formal whistleblower. I will offer a true story, and one from my own lab (but names were changed), to demonstrate how powerful informal intervention can be.

To begin the story, my style is one that values delegation, interaction, and sharing ideas, the blame, and the glory. Since I started teaching research ethics, I've told my students and staff to ask plenty of questions and to feel free to discuss ethical considerations. I want us to be on the same page in this regards. Students and postdocs have asked me about various scenarios and situations – to find the ethics in them all. I feel some sense of professional pride in being able to answer in the affirmative or to review how we do things to make governing a situation more ethical. Still, some students are not absolutely comfortable in asking me questions of an ethical nature, especially if they think I might construe questions with confrontation. I understand the potential awkwardness.

"Jill" was a new masters student in my group who had just determined the direction of her research. It focused on the overexpression of a plant gene (P) in the bioenergy crop switchgrass using a biotechnology approach with the goal of drastically increasing biomass productivity. It garnered interest of other people in the lab, including two postdocs. The postdocs started low-level efforts to help Jill and they were important internal consultants for the project. Together, we were all excited about the potential impacts of this project and eager to see Jill and the research succeed. At about the same time, this informal group was coalescing, a request for proposals was issued from a funding agency. Various people in my group wrote preproposals, but it was the P gene overexpression idea that proved to be intriguing to the agency and of my three preproposals they chose Jill's project. Jill wrote perhaps half of the preproposal and very little of the subsequent full proposal – after all, she was just a beginning masters student and didn't have the needed experience to write the full proposal. Therefore, I asked the two postdocs to take the lead as co-PIs in the proposal and work with Jill to draft the full research narrative. To make our proposal as competitive as possible they intently collected preliminary data, including the attempted cloning of the P gene. In fact, Jill had almost cloned P several times but had observed mutations in the sequence. Nonetheless, the postdocs felt she was on the brink of success, decided to help make it happen, and in a draft of the proposal had indicated the gene had indeed been cloned. For my part, I didn't know the gene had not been cloned. After reviewing the complete draft of the proposal about a week before it was due, I routed the proposal to all the outside co-PIs and collaborators, as well as the two postdocs and Jill. Coincidentally, Jill was a student in my ethics course and we had just discussed the University of Wisconsin case in class in which false data had been included on a grant proposal. The current situation was too close for comfort for her. At this juncture, what were Jill's choices? She could have instigated formal complaints against me and/or the postdocs for stating that P had been cloned when indeed it had not (even though they expected to be completed "any day now"). She could have ignored the apparent falsification, which would be unethical. Instead, she did the wisest and best thing of all. She emailed me and copied the postdocs indicating her discomfort is stating that gene P had been cloned on the proposal when, in fact, the correct sequence had not yet been cloned. This was probably a difficult email to send in that the gene probably should have been cloned a few weeks earlier, but there were reoccurring problems that were hoped to have already been rectified. As it turned out, the postdocs were truly hoping for the best, expecting imminent success and not intending dishonesty. And I was slightly out of touch. This incident provided a point of opportunity to discuss professional ethics of proposal writing and helped me to get in touch with the true progress in the project. It also helped the postdocs understand limits of wishful thinking as it pertained to proposals. We changed the line in the proposal from "had cloned gene P" to "in the process of cloning gene P." The latter verbiage was not quite as powerful as the former, but expressed the honest truth. The story ended well. We remained on the same team and friends, Jill did indeed finally cloned P soon after the proposal was submitted, and the grant was funded.

Judge yourself

✓ If you were in Jill's shoes, how comfortable would you be in confronting your mentor and postdoc colleagues?
✓ Are you amenable for others to confront or correct you?

Cultivating a culture of openness, integrity, and accountability

This chapter is written from the perspective of how integrity applies within lab groups, but certainly there is nothing wrong with extending some of these principles further out, such as to departments, disciplines, etc. Each lab has its own culture, and I'm convinced that the faculty member is a primary influencer of culture. I've seen labs in which the lab door is kept shut and upon knocking on the door someone inside cracks the door only slightly in response ("what do you want?"): a culture of secrecy. Secretive professors who are paranoid that someone wants to steal their research begets secretive lab members. A party professor often has a party lab. An intense professor has an intense lab. No matter what the style, there should be research benefits to cultivating openness, integrity, and accountability. How can this be accomplished? First, there must be the cultivation of trust among lab members and the boss so that everyone is working toward the same goal, namely to produce the best science possible. The graduate students and postdocs have additional goals to graduate and get great permanent jobs, respectively. The professor wants to have tenure and to be promoted. Everyone has personal goals for improvement. However, most people value being informed and enjoy open and frank communications. Most people also value fairness. Therefore, discussing ethical issues in the form of questions is powerful. Answering questions in an open fashion is empowering to everyone involved.

Openness in labs can also be encouraged in other ways. For example, why not put draft manuscripts and completed papers and proposals on a common computer where they can be accessed by all lab members, who then could provide comments and ask questions. Lab meetings in which everyone is valued and free to provide constructive criticism can also foster a culture of openness. Scientists must be willing to be held accountable for results, lack of results, and be expected to practice professional conduct. This all seems to start at the laboratory level and with the PI. I think that a culture of openness and accountability encourages informal complaints and dialogs to ensue within a group if a lab mate begins to go down the wrong road. Maybe intervention happens without the boss even knowing about it. Indeed, a study of within-lab intervention showed that this practice can be effective (Koocher and Keith-Spiegel 2010). In cases when a lab mate performed informal intervention by confronting another lab mate, the problem was corrected over 25% of the time. This rate might not seem very high, but consider the correction rate if no intervention had been performed. This is especially important considering that the intervener felt no negative fallout in over 40% of the cases and was treated with disrespect in only about 10% of the cases (Koocher and Keith-Spiegel 2010). Thus, informal intervention is constructive, especially as it leads to better accountability. Such accountability will, in turn, lead to higher integrity and better science, and we all hope, a decreased need for formal whistleblowing where there are no winners.

Summary

✓ Whistleblowing sometimes must be done to ensure scientific integrity and to correct large problems in research.
✓ It is best to systematically follow a number of steps before instigating a formal complaint; the costs of reporting should be counted.
✓ Many research issues can be handled early on and informally, which creates a climate of integrity.

Chapter 8

Publication Ethics of Authorship: Who Is an Author on a Scientific Paper and Why

---

ABOUT THIS CHAPTER

- Scientific publishing is the main vehicle for publicly conveying scientific results.
- Publication ethics encompasses many topics including authorship.
- Authorship can be a complicated and divisive issue.
- There are several guidelines for assigning authorship according to the contributions.
- To avoid disagreements when submitting papers for publication, it is best to discuss authorship in the early stages of a study if possible.

The paradigm of publish or perish is alive and well in academia and has matured to also consider journal impact factors, citations, and various productivity metrics such as the H-index as means to assess authorship and scientific success. And, for good reason, peer-reviewed papers in scholarly journals remain foundational to science itself. Until a result is actually published it doesn't really exist in the scientific canon. Multiauthored works, sometimes including dozens of authors, are growing in popularity, owing, partially at least, to expansive collaborations in science (Bordons and Gomez 2000) and the growth of multidisciplinary projects (Wuchty et al. 2007).

What are the general issues surrounding ethics of publishing and authorship? How important is transparency in publishing, including

Research Ethics for Scientists: A Companion for Students, Second Edition. C. Neal Stewart, Jr.
© 2023 John Wiley & Sons Ltd. Published 2023 by John Wiley & Sons Ltd.
Companion website: www.wiley.com/go/stewart/researchethics2

peer review and how decisions are made? How should allegations of research misconduct be handled by journals? How should journals be managed? How should data be made public? What about post-peer review and post-publication processes? What constitutes conflicts of interest and how should these be handled by editors and peer reviewers? These are areas of interest for a lot of people in publishing, including the Committee on Publication Ethics (www. publicationethics.org). COPE was established in 1997 by a group of editors to determine the best practices in publishing and to produce a code of conduct. COPE is a resource for anyone interested in publication ethics. Many journals and publishers have adopted COPE's practices. In particular, COPE's flow charts and case studies are useful to the community.

Since many of the topics above are included in other chapters, this chapter is mainly about authorship. Whose names should be included in the author list and in what order? Who gets to claim the coveted first authorship position? What is the best practice in this regard? Many journals have guidelines to address these questions, but there is wide variation of adoption and philosophies among authors. While some people have attempted to make this issue totally objective and formulaic, many senior authors have their own systems. Indeed, there is a variety of systems (Wilson 2002) and the underlying philosophy will be discussed here as well as tips for mediating authorship disputes. The most important component of any authorship system is thorough communication with all potential authors and collaborators early in the research process. It is important that some level of understanding and agreement is in place among lab mates and authors. Indeed, there seems to be no shortage of authorship disputes in the scientific community.

The importance of the scientific publication

Until a scientific result is peer reviewed and published in a journal, it is generally not held sacrosanct by other scientists or the public. As Macrina (2005) says, "Science benefits society only insofar as its findings are made public and applied" (p. 83). In fact, a good way to find trouble as a scientist is to bypass peer-reviewed publications and release scientific findings via television, press releases and other venues that are less than rigorous. Publication in peer-reviewed journals is the long-held conservative system that, while imperfect, is the de facto accepted method for releasing scientific results. It is only because of this fact that people who value research productivity have installed a publish-or-perish culture of rewards. Again, Macrina (2005) reiterates that scientists have a moral obligation to publish.

While scientists working in companies may not publish their research as much as university researchers (although this is changing), it is generally accepted that faculty members in most colleges and universities are expected to publish their findings regularly; it is a nearly ubiquitous requirement for getting tenured and promoted. Winning grants, receiving pay raises, scientific prizes, and election to esteemed scientific societies are typically commensurate with the number and quality of scientific publications. Bibliometric indicators of research productivity are increasingly being noticed by administrators and other people who want to quantify research impacts (Bordons and Gomez 2000; Van Noorden 2010). During my tenure-track years in the 1990s, I don't recall anyone ever mentioning citations of papers as being important to research success. Now, not only do we pay more attention to how many times a paper is cited in other papers, but there has sprung up a cottage industry of research metrics. Most indices and metrics are derivatives of numbers of papers, numbers of citations per paper, and years publishing. An updated list of the most popular metrics along with software for computing them can be found at Harzing's Publish or Perish (http://www.harzing.com/pop.htm).

Perhaps the most popular metric is Hirsch's H-index, which is defined as the number of H papers cited H times (Hirsch 2005). Let's say I author 100 papers during my career and 20 of those papers were cited 20 times or more, my H-index would be 20. Another metric, which also factors-in the number of years publishing is m. M is defined as H divided by the number of years since my first article was published. So, let's say I've been publishing for 15 years and have an H-index of 20. My m-index would be 1.3. Of course, the higher the number the better, with the highest reasonable H and m being dependent on the area of science among other factors. In my field of plant biology, $H > 50$; $m > 2$ are very rare, even for successful senior scientists. $H = 20$; $m = 1$ is quite good for a mid-career scientists in many of the life sciences. For individual researchers, H increases during a career (it is impossible for it to decrease), while m, being integrative, can go up or down. There seem to be new bibliographic indices being created every year. The point is that numbers of publication citations are becoming increasingly important in many respects, and it is vital for new scientists to understand the games and rules of measuring research productivity (even though there is no consensus that the rules are fair or even that the correct measurements are plied).

For many of these bibliographic metrics, author order is not a factor, but I think that it might become more important in the future. It only makes sense; the first author of a paper is generally regarded as the person who has done most of the work on the project and who had primary responsibility of writing the paper. In biology, the last author is typically regarded as the person whose lab the research was performed in and the primary person responsible for the research program. Often there are other authors besides these, and herein lies much controversy: the reason this chapter is included. In the "old days" there were many fewer authors on papers, and hence, fewer authorship controversies. The more we know and need to know in science, the more people it takes to do the research to take us to the next level of understanding (Wuchty et al. 2007).

Predatory publishing

"Regular" journals – both open access journals and traditional subscription-based journals – that follow the time-trusted tradition of editors securing peer reviews for submitted manuscripts have now been joined by journals and publishers who cheat the system. While predatory publishing is somewhat ill-defined (Grudniewicz et al. 2019), these "fake journals" are parasites on scholarly publishing. They will print author-contributed papers as well as collect payments from authors, but fail to perform quality-control services such as peer review. These predatory publishers rode the wave of open-access publishing for making pay-to-play schemes meant to imitate bona fide journals. Predatory publishers began appearing in large numbers in the 2000s and inspired a University of Colorado-Denver librarian – Jeffrey Beall – to create a "blacklist" of bad players: "Beall's list." In a herculean effort, he created the list beginning in 2008 and continued to curate the rapidly growing list of journals and publishers through early 2017. Various legal and institutional pressures led him to relinquish cataloging the so-called predatory publishers to various other scholars who made lists by committee. In my opinion, Beall's list played an important role in calling out bad players, but was controversial in that one person was "judge and jury" of which bin (legit or predatory) a journal belonged. And of course, there were gray zones between "normal" journals and predatory ones. The clear-cut predatory journals never actually practiced peer review. They solicited manuscripts from unsuspecting scholars, summarily published the manuscripts in short order as submitted, and then collected the fee for publishing. Some fun-loving authors wrote nonsense papers that was accepted and published by these journals. I don't receive nearly the number of invitations from predatory publishers as I did a few years ago, so perhaps this problem is waning.

Judge yourself

✓ How do you feel about credit and fairness of credit assignment?
✓ How do you feel about being judged using metrics?
✓ Do you like to write or just do experiments and play a support role?

Who should be listed as an author on a scientific paper?

During the research ethics course I co-teach, we ask students to report to the group about the instructions for authors in some of their favorite journals; of interest is the journal's criteria for author-ship. Journal officials and scientists agree that it is imperative that the most appropriate people are listed as authors on a paper on the basis of contributions: both on the research and in writing the manuscript. Important also is that authors appear in a logical and fair order, again, because of their roles on the study and manuscript. For each of these items, there is apparently no universal formula agreed upon by everyone to distribute fair credit. Except for the rare journal that does not address authorship at all, journals fall into two camps. The first camp uses the words "substantial contribution" and a few qualifiers to delineate who should be included in the authorship list and who should be relegated to the acknowledgments section. This first camp is pretty ambiguous in that responsibility falls to the lead authors – especially the corresponding author – to determine who should be an author by virtue of some subjective sub-criteria. The overriding "substantial contribution" criterion is purposely vague, which gives senior authors a lot of leeway to make decisions. Many editors and scientists don't want to be shoehorned into a formula. The second camp invokes guidelines developed by the International Committee of Medical Journal Editors (ICMJE). See https://www.icmje.org/recommendations/browse/roles-and-responsibilities/defining-the-role-of-authors-and-contributors.html for specifics.

ICMJE has established four requirements for authorship: 1) Substantial contributions; 2) Writing and intellectual contributions in sculpting the manuscript; 3) Approval of the paper; 4) Accept responsibility for the content.

1. The author made substantial contribution to the study. The first requirement can be fulfilled in one of three ways. An author would have made substantial contributions to: 1) the conception and design of the research, or 2) collection of data, or 3) data analysis or interpretation of data. If someone is intricately involved in a study, meeting one of these first sub-requirements is typically not difficult, but again relies on the interpretations of what "substantial contributions" really means. But at least ICMJE has sought to clarify the definition of roles of "substantial contributions."
2. The author drafting the manuscript and/or providing intellectual contributions on revisions. This second step is often shortchanged, thereby seemingly disqualifying many potential authors from authorship if ICMJE guidelines are to be taken literally. Here is the reason. I can't recall how many manuscripts I've sent out to co-authors who subsequently never make any revisions to the draft of the paper. "Looks great," they say, if they return my emails at all. And then there is the case where primary author(s) never vets the drafts to their co-authors. If authors don't see the manuscript, this makes the third ICMJE requirement also impossible to meet.
3. An author gives approval to the paper to be published. How many researchers, especially those who publish in journals that adopt the ICMJE criteria take these criteria seriously when assigning credit in authorship? Glenn McGee (2007) suggests that the answer is few. He cites a study in which only three of ten non-corresponding authors met the ICMJE guidelines for authorship. He also cites problems of ghost authorship. In scientific writing, ghost authorship companies or other authors ask a prominent scientist to agree to being an author on a paper in which he or she had no involvement as a means to boost a paper's credibility or

visibility. See the case study at the end of this chapter for another kind of example of ghost authorship. Again, the lead authors (first and corresponding authors should make sure that authors have seen and approve of the manuscript when submitted and the final paper to be published.

4. An author agrees to be held accountable for the work. This criterion is especially important if there is an investigation and/or retraction of the paper. That willingness to be responsible for the content of the paper is important, but in my experience, not all authors of a paper can be held to the same standard of accountability. It depends on the role of an author, experience, and depth and breadth of expertise. One would not expect an undergraduate student who performed an isolated experiment and analyzed only his/her data to have the same responsibility as the postdoc first author who presumably had knowledge over most, if not all, of the study.

Providing funding or simply being the boss is not grounds for authorship (Macrina 2005). In addition, providing equipment or lab space also does not qualify scientists for authorship (Macrina 2005). I know of senior scientists who pad their resumes by this means and it is unethical. Likewise, being the head of department or center also does not warrant authorship. Indeed, author inflation seems to be rampant, with many, many authors appearing on papers these days that is likely unwarranted. McGee (2007) believes that education is an important tool to address the practice of author inflation.

There is another issue that is at least, if not more, damaging when assessing contributorship is assuring that people who meet author-ship criteria are not excluded on the basis of status or other reasons. This, too, is unethical. "Roger Croll (1984) in his landmark paper on the topic emphasizes that hourly wages, academic credit, salary, and commission are irrelevant in assessing credit. This is counter to the rationale sometimes advanced as to why technicians, consultants, and others should not be listed as authors, that is, that they have been compensated for their contribution. This argument ignores the fact that generally all who contribute to the project are paid to do so, directly or indirectly" (Pool 1997, p. 129). Indeed, there are certain organizations, such as the United States Department of Agriculture-Agricultural Research Service (USDA-ARS), which, by policy, historically excluded technicians as authors. To include a technician as an author, it would have to be approved by the ARS Research Leader, Centre Director or Location Director (ARS 152.2 amended in 1997), while the policy acknowledged that if a technician performed the work of an author, it is unethical to deny authorship. (Note that the USDA-ARS policies are now behind a "sign-in wall" and no longer available for open access reading.) Perhaps the agency has modernized their practices even if they don't advertise it.

Why is it important to correctly list only the authors who substantially participate in a research project? One reason applies to most of the things we do in science: to get it right by assigning credit to those whom credit is due. As seen above, there are two kinds of common authorship mistakes. The first is not listing authors who should be on a paper. Leaving people off the authorship list when they should be listed can cause hard feelings and a sense of injustice and unfairness, even if it is an organizational policy such as one that existed in the USDA-ARS. Everyone knows that unjust exclusionary practices are not good for morale or productivity. Science ought to be transparent. The second mistake, as you might have guessed, is including people as authors who are not qualified to be authors on a particular paper. Some people may believe that listing lots of authors makes a paper seem more important. Alternatively, some people may think that listing too many authors dilutes the "real" authorship. I have a difficult time understanding these perceived effects. Perhaps a bigger problem is authorship coercion to include people who shouldn't be included as authors as a quid pro quo move. See the case study at the end of the chapter that deals with this. Just because a faculty member is on a graduate committee, gives some materials to the student, or lets a student use a piece of equipment that does not give the faculty member the right to coerce authorship from a junior faculty member or student. Corresponding authors ought to monitor these issues. Judging from stories I hear and my own observations, I judge that this bullying/coercion is a much more serious and broader issue in practice. Such bullying is not always professor-on-student. I've observed postdocs bully other postdocs and technicians, with one desired effect being to strike a person from authorship roles and/or remove a person from a more prestigious author spot to a less prestigious spot. I've observed such reward/punishment attempts with shock of the audacity of the perpetrator. It is serious because it gives the appearance that someone is contributing more than he or she really has. It also fosters ill-morale, and it causes downstream injustices with regards to both students and faculty members with regards to resource allocation and stature. I've heard an administrator express the concern,

"I'm not sure that some of my faculty even have research programmes, but I see their names on lots of papers. What does that mean?"

Because of these issues, I've become a big fan of journals requiring a section that lists author contributions to a paper. The list should be sufficient to denote unambiguous roles in the study and manuscript. In most cases, the contributorship should map back to the ICMJE criteria. As a reader of a paper, it also helps me to contact the appropriate person should I have any questions about certain details described in a paper. During tenure decisions or similar, a committee can see exactly what a person's contribution was to any specific paper. Even more importantly, should a paper come under investigation or critical scrutiny, the author contribution statement may indicate who should be named as a respondent and who may be eliminated as a respondent.

In late 2022, the first instance of authorship attribution was given to an artificial intelligence (AI) app. In this case ChatGPT – a large language model – was listed as an author on a preprint deposited in medRxiv. As of early 2023, ChatGPT has gained notoriety in composing several works. Scientific publishers quickly responded by placing human authors on notice that ChatGPT (as well as other large language model apps such as Google's Bard) don't meet the criteria for authorship and therefore shouldn't be listed as authors. For instance, ChatGPT can't be held accountable as an author of a scientific paper, e.g., ICMJE criterion #4. Indeed, the development of such AI has the potential to disrupt scientific publication by its mere presence with potential use and abuse. As this book was entering production, ChatGPT and Bard were just becoming widely used. Some people envisage these large language models as being helpful in legitimate scientific writing, while others are calling for them to be banned in institutions and all academic works. By the time this book is published, the debate will continue to rage with likely no consensus as to the legitimate utility of large language models in scientific publishing.

Judge yourself

✓ How important is it to assure that authorship in your science is assigned correctly? How strongly do you hold your beliefs and how hard will you fight for appropriate authorship attribution – especially for others?

✓ What are some ways you can influence best practices in authorship?

✓ Is it worse to include an author without good reason or exclude someone who should be an author?

✓ Do you prefer loose or stringent criteria for inclusion in authorship? Why?

How to avoid authorship quandaries and disputes

First of all, one cannot over-emphasize or over-discuss authorship or authorship order during the course of a research project and while a paper is being drafted (unless, of course, an unscrupulous author is trying to lobby to boot or demote another author via politicking. Indeed, I've observed bullying in this regard.) In fact, delineation of authorship philosophy can begin when a grant proposal is being written, when a visiting scientist or graduate student joins the lab, during lab meetings, or even as graduate student and postdoc candidates are being interviewed. Certainly, an entering prospective lab member should have an understanding of the publishing culture of a PI and lab. I tell everyone in my group that I want them to "own" their studies and be the first author of the research they drive, but to include others if it is mutually beneficial and appropriate. My philosophy can be summed up as "the more the merrier if warranted." I tell them to then put their co-authors to work in not only the experimental areas, but also in writing and revision (the ICMJE second requirement), but to not take advantage of people. I also advise them to put my name as the last author if they wish and put me to work too. I typically delegate to the first author the duties of selecting their co-authors ("and make sure you don't omit anyone") and then I'll discuss with the first author about any adjustments that I believe need to be made. I really appreciate having consensus among authors of a paper on the authorship list and order, but unfortunately, on occasion, that doesn't happen.

My style is not the only one that works and it is important to think about what your own style will be as a PI. A lab-wide system is important, and communicating the system is important to avoid problems that can eat up time and strain relationships. Let's imagine there are three graduate students performing related research in a lab. They all work together and understand that each one will be first author of their respective main projects. If one of them (say, Adam) decides to exclude the other two students from authoring his paper even when the others included Adam as an author on their papers, hard feelings will ensue if they've all met an agreed set of criteria, such as the ICMJE guidelines. These kinds of scenarios are not pretty. It is better to be inclusive (yet make your co-authors earn their keep) than to be too exclusive in authorship.

There are other authorship issues that must be negotiated. Authorship order, the journal for a submission, when to submit, and the scope of the study to be published are all items that could engender points of disagreement among authors. In some labs, the PI simply dictates the details. In other labs, the decisions are more free-flowing and negotiated among authors. It might be that one person wants to publish lots of small papers and another person wishes to pool reams of data into a large comprehensive study (with that person as first author!). This decision is actually an important point when one considers citations and "fame." Only one person can be the first author on the paper (even when it is indicated that "the first two authors made equal contributions to the paper") − everyone else, including the corresponding author or PI is "et al." in a citation when there are more than two authors. Who is listed as corresponding author is important and is sometimes negotiated among authors. Personally, as the PI, I don't care if I'm corresponding author (the one who submits, deals with revisions and page proofs, and the one who pays page charges or open access fee), but I'm told that some people look carefully at this as a sign of independence and being the PI. Some people consider the corresponding author to be the "brains" of a paper. And so it is appropriate if someone other than the PI be the corresponding author if appropriate, i.e., if that person is doing the work of a corresponding author. It is smart to be savvy and discuss these issues up front to avoid troubles when the paper is submitted.

It is important to reiterate that the fourth ICMJE criterion is important: that authorship is tied to responsibility and accountability. It is assumed that authors should be directly responsible for the overall content of a paper, whereas certain authors on multi-authored papers are responsible for specific parts of the paper. It is important to consider that your name as an author will be associated with other co-authors on both the wonderful and terrible papers. Being an author on a "paper of the year" is wonderful and being an author on a badly flawed and retracted paper is terrible. Guilt by association happens. There is probably some disagreement about which authors are directly responsible when something goes wrong. I can envision situations in which not every author may be directly responsible and accountable for mistakes, but again, we come back to guilt by association. Responsibility and accountability is worth discussing with lab mates and co-authors as research and papers are developed. It is important to understand what your collaborators and lab mates believe about these issues. This whole business demonstrates the importance of choosing wisely and taking seriously the entirety of each paper and the details found within.

Authorship for works other than research papers

We have delved into the rules of authorship for research papers that are to be published in peer-reviewed journals – these are the most crucial of publications. Not discussed yet are rules for conference abstracts, book chapters, and review papers. In addition, it is important in some settings to discuss inventorship criteria for patents. Review papers are often invited works that review a field of study and published in a peer-reviewed journal. Book chapters are typically similar in scope and purpose and may or may not be peer-reviewed. For either reviews or book chapters, the main author invited to contribute might invite other authors to participate in writing the piece. It is assumed that all authors play important and crucial roles to the papers, but there are no rules similar to those found in ICMJE. Conference abstracts also don't seem to have authorship rules. In many ways, abstracts and the ensuing posters or oral presentations are harbingers of future peer-reviewed publications. With that in mind, authors tend to include all the future authors of a paper, but perhaps with the order slightly altered. Thus, authorship arrangement for conference abstracts is good "practice" for preparing the authorship list and order for downstream papers. For some fields it is typical that the first author is actually giving the talk or presenting the poster, when in practice, he or she might appear somewhere else in the author order in the actual paper. The real problem often occurs in that not as much thought or consideration is usually given to abstracts compared with the full paper that might follow. Sometimes a deadline sneaks up on the chief abstract author. I'm quite sure that I'm an author on abstracts I've never seen, and I've placed authors on my own abstracts in the same boat. Sorry about that! In reality, professionalism dictates that each author should see and agree with the contents of conference abstracts even though it is unlikely that all ICMJE criteria can be met at the time an abstract is submitted because of timing and logistics. For each of these types of publications, it is good to avoid authorship surprises at the end. Communication is vital.

Paper mills

In publishing, a paper mill is a company that fabricates studies and papers, which is the product they sell to "authors." Some paper mills trade in authorship slots on legitimate studies/ manuscripts. Thus, a paper mill client may either buy or sell authorship slots and the paper mill is the broker. On many occasions, the entire product is made up and fraudulent. There have been thousands of papers published from paper mills, and in some cases, paper mill companies have targeted specific journals. In some extreme cases, there may have been purported collusion with leaders within the publishing operation, e.g., editors. Journals and publishers have found several items in common with paper mill papers. These include fake western blot images and other gel images with features that are not intrinsic to real blot/gel images. Similar-looking plots in papers that are not related, as well as similar-themed articles from different authors (Else and Van Noorden 2021). That is to say, unrelated papers appear to be related: that is because they were manufactured at the same paper mill.

The difference between authorship on scientific papers and inventorship on patents

Who is listed as an inventor on a patent is governed by patent law. Just because someone is an author on a paper that describes an invention does not mean that the same person can legally be listed as an inventor. While authors can be determined using the ICMJE guidelines, inventors are people who play a critical role to the conception of an invention. In fact a US court case, *University of Pittsburgh v. Hedrick*, demonstrated the importance of participation in the conception of an invention (rather than showing that an invention really works) to inventorship; it is critical. Warren and Cao (2009) conclude that the legal finding on the Hedrick case demonstrates that inventors need not know that an invention will work, but that the crucial aspect is that they have all the details of the invention worked out in their minds, which is backed up by lab notebooks or other written evidence. The Hedrick case seemingly contradicts common sense ethics discussed on publication authorship and is therefore worth discussing.

University of Pittsburgh researcher Adam Katz and Ramon Llull conceived of a method to isolate stem cells from fat cells, which could then be differentiated into muscle cells or bone cells, or other cell types. A UCLA researcher, Marc Hedrick, joined the group at Pitt for a research fellowship for one year. When Katz and Llull filed a university disclosure, the precursor for a patent application, Hedrick was included as an inventor. Indeed, Hedrick continued his part of the research after he returned to back to UCLA and showed that the invention actually worked. Subsequently the patent was issued. But then, the University of Pittsburgh filed suit to remove Hedrick from the issued patent, and they won. Warren and Cao (2009) sum that, "Even though Drs. Katz' and Llull's work was scientifically inconclusive at the time of conception, the later evidence providing scientific certainty was merely a reduction to practice and not an inventive contribution." Counterintuitive with regards to publication authorship ethics, trying to include people who helped in an invention but do not fit the legal description of an inventor could lead to hard feelings and lawsuits.

Other thoughts on authorship and publications

It is especially important for young scientists to aggressively publish, be collaborative, and increase our collective knowledge in science. Most institutions value independent research programs that yield good science as evidenced by contributions on publications. I don't object to the trend toward valuing bibliographic metrics based upon the numbers of papers authored and cited. Indeed, these help me see what research is valued and used by others. I encourage my trainees to frame their papers in such a way to garner to most citations possible, which indicates that the information is being used and spread – that it is relevant to science and society. There is a fine line between a noble approach of sharing knowledge that also acknowledges career development strategies vs. doing science to get the paper published at any cost. A career is not built upon a single paper. Collegiality, sustained effort, and best ethical practices are vital considerations in the long haul over a career.

Because of bibliometrics and career considerations I think it is becoming crucial to be able to identify an author unambiguously. There are efforts to assign a unique identifying number to each author. The predominant system, which is increasingly becoming mandatory as an author for some journals is the ORCID (Open Researcher and Contributor ID). My ORCID is 0000-0003-3026-9193. I like the idea of establishing a central numbering/tracking system for authorship for several reasons. Other than having the ability for unambiguous tracking of authorship, such a system could be helpful in evaluating productivity of a known or unknown scientist – maybe someone wishing to collaborate with you (to answer for example, "How much does he or she publish?"), or known scientists coming up for promotion to full professor (e.g., "How much do their peers cite their works?"). Another very nice tool in that respect is the Google Scholar Profile, which is widely adopted by scholars, wherein Google lists an author's papers, how many times they've been cited, and by whom, as well as bibliographic metrics such as the H-index. As an author, Google Scholar Profile also provides a convenient listing of my papers for the world to see.

I recommend all my trainees take an ORCID – the earlier the better. I also recommend that students and beginning scientists who have very common names take a pen name that is unique to science and use it faithfully with absolutely no variation. For example, I'm one of a few "C.N. Stewarts" around in science, but there are more X. Zhangs and D. Patels than I can count. Popular surnames work against unique individuality in tabulating bibliometrics. But if a Xin Zhang were to add two more initials, say Y and Z, then he would have a much more unambiguous name in science, especially in any particular field. There is one "XYZ Zhang" that I found in Harzing's Publish or Perish and just one "DCB Patel." This strategy worked for US President Harry S. Truman, who, in fact, had no middle name. Adding a few novel initials to an author's name to make it unique should make for effective identifiers. If this practice is adopted, the author should always include all initials in every work. ORCID does the same thing, but by assigning each author a 14-digit number. Do both! Because "branding" is so important these days for scientists – essentially for other scientists to associate a name with a field of science – I also advise my trainees to pick a name and use it throughout their careers to make sure it is easy to identify individual scientists. But there seems to be a movement counter to this advice, which is to retrospectively allow name changes on published papers. Here's one way it would work: let's say a fictional researcher named, say, Bruce Jenner publishes research in track and field kinesiology and becomes well known in the field, even winning various awards and medals. After transgendering to become known as Caitlyn Jenner, this author now would not only like to publish under the name "Caitlyn Jenner," but would like to retrospectively change "Bruce Jenner" to "Caitlyn Jenner" in all previously published papers. In 2021 the publishers Springer Nature and Wiley now allow retrospective name changes to occur. I suppose if a scientist was very famous, then it would be widespread knowledge that Bruce and Caitlyn are the same person, but I could see that others might be confused. And then there would be the confusion in citing the papers with the apparent name changes, correcting the scientific record, and many other things I haven't begun to think about. I suppose this is another compelling reason why each scientist should have an identifying number as well: to keep track of potential name changes that could occur someday in the future.

Judge yourself

✓ How do you feel about being responsible, as an author, for a large multidisciplinary project? Accountable?
✓ How important is it to trust your collaborators and co-authors?
✓ How important is it to be associated with "winners?"
✓ How do you feel about your name and its intrinsic value? Would you mind taking a pen name?
✓ How would you feel about "becoming a number" a la ORCID?

Case study 1: who is an author?

This case study is reprinted courtesy of Daniel Vasgird, PhD, and Ruth L. Fischbach, PhD, MPE, and the Trustees of Columbia University in the City of New York, http://www.ccnmtl.columbia.edu/projects/rcr

Susan Jacobs, a PhD student from a small university, sets up, as part of finishing her dissertation, a six-month internship at a prestigious larger institution in order to learn a new molecular-biological technique. Ms. Jacobs contacted the laboratory leader, Dr. Marvin Frank, a world-renowned scientist, in the hope of developing new skills for her research and also to foster a relationship with Dr. Frank, who is well connected in her field of biochemistry.

When Ms. Jacobs comes to Dr. Frank's laboratory, she is greeted warmly as a member of the team. Dr. Frank, the graduate students, the postdoctoral fellows, and the technicians include Ms. Jacobs in the weekly laboratory meetings, in which everyone participates in a free exchange of ideas about the ongoing projects in the laboratory, and which last for hours. In the meetings, Ms. Jacobs finds some of the ideas helpful but others less so and gives her point of view concerning the ongoing projects. In addition, she meets weekly, one on one, with Dr. Frank, who provides significant scientific advice and one or two recommendations, which advance her work and move her in a slightly different direction. She discusses the results of her research with her mentor, Dr. Melissa Seabrook, back at her home college, by weekly e-mails and occasional phone calls, interactions that also push ahead the project she started in Dr. Seabrook's lab three years ago.

Ms. Jacobs makes great progress during the six months she spends in Dr. Frank's laboratory, and she writes a paper reflecting some important findings. Ms. Jacobs puts herself down as first author, Dr. Frank as second author, and Dr. Seabrook as last author on the paper. At the end of the paper, she gives an acknowledgment to a technician who showed her several techniques and worked with her on a few experiments.

Ms. Jacobs based her listing of authors on her understanding of the guidelines put forth by the International Committee of Medical Journal Editors (ICMJE), which say that an author is someone who has made significant contributions to the conception and design, or to the acquisition of data, or to the analysis and interpretation of data; was involved in drafting the article or revising it critically for important intellectual content; provided final approval of the version to be published; agreeing to be held accountable. The guidelines, which are followed by approximately 500 medical journals, say that all four criteria must be met for authorship. Ms. Jacobs would like to send her manuscript to a journal that follows ICMJE guidelines as soon as possible, because of what she feels is the importance of her results.

Ms. Jacobs gives Dr. Frank and Dr. Seabrook a draft of her manuscript for review on a Friday, hoping for feedback by Monday. Dr. Seabrook sends her comments by email to Ms. Jacobs. Dr. Frank sends his comments back to Ms. Jacobs and changes the authorship listing to include Ms. Jacobs, the technician, two postdocs in his lab, two graduate students in the lab, himself, and Dr. Seabrook. Dr. Frank also gives a copy of the draft to all the members of his laboratory for discussion at the next meeting.

Ms. Jacobs is shocked that Dr. Frank added the other laboratory members to the draft, explaining to him the ICMJE guidelines and maintaining that the major intellectual and physical work in preparing the paper was done by her and by Dr. Seabrook and Dr. Frank. Dr. Frank is equally surprised by Ms. Jacobs's feelings, responding that he and Ms. Jacobs benefited from the input of all the other lab members. Dr. Frank adds that a graduate student in the laboratory, Lisa Bain, is writing a short paper that is based on some very exciting preliminary findings, and that Ms. Jacobs would be included in the list of authors. Dr. Frank says that the results of Ms. Bain's research would need further elaboration in the laboratory and that a second paper using the same data and additional studies would be more comprehensive, and that Ms. Jacobs would be included on the second one, too.

Dr. Frank insists to Ms. Jacobs that the contributions of all the laboratory members were sufficient to satisfy the ICJME guidelines for both papers, adding that the idea of a scientist acting as an independent entity is an outdated concept and that those who work around a scientist contribute significantly, helping him or her to function.

Ms. Jacobs tells Dr. Frank that she does not want to be included on Ms. Bain's paper, feeling that she did not contribute adequately. Dr. Seabrook, who follows ICMJE guidelines but was intimidated by Dr. Frank's stature, advises Ms. Jacobs not to rock the boat, to use Dr. Frank's revisions and some of the changes suggested during the laboratory review and to submit the paper to the journal with the authorship he suggested.

1. Why should Ms. Jacobs and Dr. Frank have discussed the laboratory's approach to authorship issues when she started working in his laboratory?
2. Why is the order of authorship and the listing of authors important in a research paper?
3. What is the difference between an acknowledgment and a listing as an author?
4. Although many journals subscribe to the guidelines of the International Committee of Medical Journal Editors, many do not, and many researchers do not follow the practices that it recommends. What tends to happen, and how are ICMJE standards being challenged?
5. Who among the authors takes responsibility for submitting the paper to a journal and following up with the editor and peer-review revisions?
6. What are some potential problems with Dr. Frank's submitting a paper on preliminary findings and not performing sufficient corroboratory experiments?
7. What kind of problems may arise if the same data is used in multiple papers in the research literature?
8. What might happen if someone is listed as an author on a paper for which he or she did not do any work?
9. What might have been done to resolve Ms. Jacobs's ethical dilemma with Dr. Frank about the authors on the paper?

Case study 2: the case of the ghost-writing student

Courtesy of graduate student Hong S. Moon

Dr. H. Dick Dockers is an assistant professor in a tenure-track position in a department of engineering at a major research university in California. In his statistical modeling laboratory, there is one full-time graduate student, Leonard Smyth, and one part-time student, Johnston Klub. Smyth is a PhD student in the third year of his graduate study who is focusing on developing quantitative models for automotive industry. Smyth's research project is funded by one of the major car manufacturers. The part-time student, Johnston Klub, is pursuing master's degree in Dr. Dockers' laboratory while he is working at the university in the dean's office as a staff-member.

Dr. Dockers receives a phone call from the car manufacturer's external funding officer with the bad news that the company can't continue to provide research funding, since the company is experiencing financial difficulties. Dr. Dockers has been submitting several grant proposals to various funding agencies, but with no success. The car grant is his only grant. Without further funding, Dr. Dockers can't support Smyth, the full-time student. Two years after Klub joined the lab as a part-time masters student, he was promoted to be the assistant director of university's research funding office. Klub wants to graduate soon, but his day job has made him far too busy to write his thesis. Klub explains his situation to Dr. Dockers and asks him for help. Klub implies that he may be able to provide funding to Dr. Dockers laboratory if he "helps" Klub to quickly get his master's degree.

Dr. Dockers brings Smyth in his office and tells him that his funding will be terminated soon. Dr. Dockers tells Smyth he has three options. One is Smyth writing a grant proposal to get funding for his research. And Dr. Dockers says that writing a grant proposal is not easy and takes a long time. A second option is that Smyth can find another advisor and give up his research project; this option would extend the time in graduate studies another three or four years. In addition to these problems, none of the professors in automotive- related research at his university has enough funds to accept a new graduate student, since all car manufacturers are experiencing tough financial situations.

The last option is that Smyth can "help" Klub to write his master's thesis. When Smyth asks Dr. Dockers to clarify the term "help," Johnson realizes that "help" means "writing the thesis" for Klub. Smyth's reward for writing the thesis for Klub would be an arrangement for research funding to come from the university's research office for the next two years allowing

Smyth to graduate. Dr. Dockers leaves the final decision on Smyth. Although Dr. Dockers mentions that last option is probably the best vehicle for uninterrupted funding for Smyth to continue his graduate program. Smyth leaves Dr. Dockers office and starts thinking about his future.

1. What is the best solution for Smyth to continue his current research and graduate?
2. "Ghostwriting" is defined as writing in the name of another. There are many commercially operated "ghostwriting" websites. Is it ethically acceptable to publish ghostwritten scientific papers?
3. There are some science fields that acquiesce "ghostwriting." Some people argue that writing is not a part of sciences, thus ghostwriting is acceptable unlike data fabrication or stealing intellectual property. Also they say that you can be more productive if you are working in the lab instead of spending a lot of time to write a paper. What do you think about this opinion?
4. Are there any other ethical problems in this case study?

Summary

✓ Authorship is one of the most important considerations in science where positions, tenure, and promotions are at stake.

✓ There is still some intrinsic unfairness in the world of authorship where ethics are not considered.

✓ Issues of responsibility, accountability, research metrics, and keeping track of authors are all issues that are worthy of discussion and debate.

Chapter 9

Grant Proposals: Ethics and Success Intertwined

ABOUT THIS CHAPTER

- Science cannot be accomplished without funding from grants.
- Granting opportunities are increasingly competitive.
- Finding funding the right way and playing by the rules set up successful research.
- Being a fair and ethical collaborator is an important and a big part of modern science.
- Performing the research in the way proposed is a critical commitment in science.

Along with publishing, funding is one of the cornerstones of modern science. Simply, without outside funding, little research can be accomplished. While grant proposals in the United States are most often are submitted to national agencies such as the National Institutes of Health (NIH) and the National Science Foundation (NSF), there are a plethora of companies and foundations that support science. In almost all cases, savvy grantsmanship and collaboration are necessary for funding, especially for landing bigger grants. As with all chapters, but especially so here, best practices and ethical behavior pave the winning path. Case studies will focus on proposal writing and collaboration. The ethics of multiple proposal submissions and doing the work before funding occurs will also be discussed as well as financial matters. We'll examine features of being a good PI and a good collaborator.

Research Ethics for Scientists: A Companion for Students, Second Edition. C. Neal Stewart, Jr.
© 2023 John Wiley & Sons Ltd. Published 2023 by John Wiley & Sons Ltd.
Companion website: www.wiley.com/go/stewart/researchethics2

Why funding is crucial

In the olden days, there were far fewer scientists than today, many of which were wealthy in their own right and needed no outside funding. Relative to today, nineteenth century (and earlier) science was at its rudimentary stages and was pretty cheap to do (Shamoo and Resnik 2003). Check out 1960s sci-fi movie or TV shows and you'll see "futuristic" labs appearing quite sparse, and mainly featuring fancy glassware containing different colored liquid – and giant computers with tape reels. This reflected not what people thought science would be so much as what it was at the time. Other than rare "big science" projects such as the Manhattan Project to develop nuclear weapons in World War II and space-mission science in the 1960s and 1970s, most twentieth century science was relatively cheap and simple. Even 30 years ago, the entire landscape of science was vastly different, which included far fewer scientists and more available funding per scientist (Shamoo and Resnik 2003). Therefore, the competition among scientists for funding was less stringent and competitive than today (Figures 9.1 and 9.2). As an example, in 2009, the US "stimulus package" infused new funds into science. The ARRA plowed over $10 billion (USD) of one-time "stimulus funds" into the NIH budget that was open for competition by US biomedical and biological scientists. Every university administrator in the country actively urged

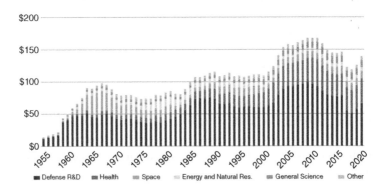

Figure 9.1   US federal research budget in billions of dollars by type of research. Note that defense research and development has the highest funding in any given year.

Source: Reproduced from https://www.bakerinstitute.org/research/us-federal-scientific-research-and-developmentbudget-overview-and-outlook Rice University's Baker Institute/with permission of Rice University's Baker Institute/last accessed December 12, 2022.

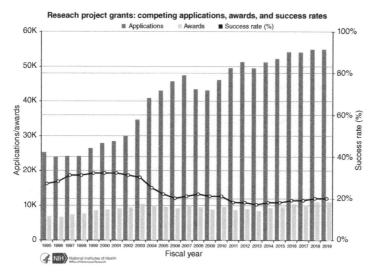

Figure 9.2 National Institutes of Health R01 grant funding rate over time.

faculty to go after this one-time pot of free money. As a result, a scientific feeding frenzy ensued resulting in about 20,000 applications (which also required an even greater number of peer reviewers) and, as a result, a paltry 1% funding rate. This, in turn, led to slowing down the wheels of review and award. Who knows how many wasted person hours were spent writing failed proposals that would never truly be competitive. But, what is the alternative – funding starvation? And I kind of doubt that the 99% or researchers who were turned away this competition simply discarded their proposals. No! They reused and recycled, submitting slightly revised proposals in subsequent competitions in hopes that odds will fall their way. It is important to note that competition for U.S. grants is already very intense given that government funding has been relatively flat (adjusted for inflation) for years but has had a number of waves (Figure 9.1). There are far more investigators requesting funds than available. For example, NIH R01 grant proposals are funded at around 20%.

Fact 1: scientific research is expensive

Salary and fringe benefits for graduate students, postdocs, technical support, and other people are expensive, especially in the developed world. Gone are the days where students of wealthy sponsors work for free.

Materials and equipment needed for modern life scientific research is not cheap. And those are just the "direct costs." On top of direct costs, universities charge funding agencies money to administer grants. This is known variably as indirect costs, facilities and administration (F&A), or overhead charges. Indirect costs are used by university to pay for utilities, clerical help, cost accounting, and for maintaining facilities. In some universities and departments, a portion of these funds are shared back to departments and PIs. The indirect cost rate varies per institution and can range from 40% of direct costs to 70% or even higher. Therefore, the tough game of getting funding is made even tougher by having to budget for the indirect costs required by the home university. The typical grant size in the United States ranges from $500,000 to $1,000,000 for many US federal agencies, which is used to fund work on a project for a duration of three or four years.

Fact 2: if a PI is not doing competitive science, then funding chances are low and little meaningful science can be accomplished

Bank robbers rob banks because that is where the money is. Scientists don't typically enter a field of science because they enjoy seeking funding, but because they want to change the world in an area of science they live. But to stay active in a field and produce meaningful data, funding is required. Therefore, it seems to me that if I had a choice of doing science in a field with little funds available or one with lots of funding possibilities and be happy either way, then why not choose the easier-to-fund field? Research productivity (and enjoyability) increases when the proposal writing effort to funding ratio is low and the funding level to administration headaches is high. Spending a lot of time writing grant proposals can be tedious. Spending a lot of time writing failed grant proposals is frustrating. Writing no grant proposals and getting no money is also deemed failure and won't be tolerated by most employers at research universities. So the middle ground is maximizing funding potential while decreasing efforts to get those funds. So, learning where the money is and how to get it is crucial to thriving in science. Some faculty vacillate from aiming too high, where they might continually fail, to a place where they aim too low and acquire low amounts of funding to do research that has little real impact. The key is for each researcher to find the funding sweet spot, which of course will vary among researchers and fields of science. For years I wrote a lot of grant proposals for standard-sized single PI grants along with cultivating relationships with companies. Then I started leading larger teams into proposals and also collaborating as a co-PI on bigger grants. These efforts take more time so I actually had fewer (but larger) awards. Someday, my preferred funding model may change yet again.

Fact 3: the faster money is spent, the more science will be accomplished

Many scientists hate writing grant proposals and therefore want to maximize the time that money is held. This strategy is a mistake. The highest budget line in grants is typically for people – salaries and fringe benefits of students and scientists who will actually do the research – and so hiring and spending (within reason) is correlated with accomplishments. Of course, there are several caveats, such as hiring the right people and making sure their funding won't run out within a reasonable time (you don't want your PhD students to run out of stipend just two years into their programs). I've seen this happen where students are either abruptly dropped or prematurely awarded a PhD by unethical PIs: a travesty. You also don't want to create anxiety among your researchers – you want them to be focused on research. Most granting agencies do allow for "no cost extensions," giving PIs extra time to do the research. But the sooner researchers can spend the money and accomplish the goals of a project, the sooner that new funds can sought for follow-up research; science should be performed in a timely fashion.

Money and ethics

Given the intensifying funding situations and the intrinsic drive for productivity that most scientists possess, many ethical dilemmas can emerge. Temptation to make ethical shortcuts is ever-present. First, getting funding every year is one of the most stressful things that scientists do. It is probably the *most* stressful issue that keeps me awake at night. Sometimes in the middle of the night, I dream of 40 hungry mouths to feed. It can be a nightmare, actually. But, the case can be made that without at least some of this stress there is no science. Second, to have sustained funding, not only must science be sound but people must be treated ethically and effectively mentored. So, the theme for this chapter is that success in funding is partly the result of ethical treatment of collaborators, subordinates, administrators, and fellow scientists. That is, science funding is positively correlated with ethical practices in grants and contracts.

Judge yourself

✓ How do you generally deal with stressful decisions that could be entangled in ethics?
✓ Are you competitive?
✓ Do you enjoy planning work and then "selling" a proposal?
✓ What's more important – people or money?

Path to success in funding

This chapter is not intended on guiding the readers toward specific grants. Rather, the focus is in offering guidance about ethical issues that are inevitable on the route to winning and spending research funds. Ignoring ethics on the former can hamper proposal writing, thwart funding, and sustained success. Spending funds improperly can land a PI in hot water with the funding agency and their institution. Certainly, FFP does not spell success when it comes to acquiring sustained funding.

Essentially, successful grant getters have several common attributes. First, they all have great ideas. Second, they are able to convey those ideas and the benefits of funding them to a group of peers (for example, see the Heilmeier Catechism in the box below). Third, they realize that success follows success (i.e., funding follows publications, leading to more of both). Fourth, people trust them with resources (money) and have faith that PIs will do what they say they'll do. Most grant proposals are formally peer-reviewed. However, in some agencies, program managers or officers have a lot of leeway to fund proposals they want funded. And inevitably, reviewers and program managers tend to think of funding PIs rather than projects, even though proposals are project-centered (and most are). This is an important consideration. It is human nature to think in terms of funding people. Program managers also exist (agencies may call them by various titles) in many companies and private foundations. There are even private individuals who endow or otherwise fund research at a university. One trick of getting funded is matching great ideas with people and organizations who want to fund those great ideas. Therefore, a smart scientist will think in terms of matchmaking and goodness-of-fit to maximize research efforts and minimize frustration. Great ideas can actually carry the day, especially for young scientists with little track record of research success. Be bold in your ideas! But after that first grant is awarded and the money is spent, nearly everyone wants to know how much good science was done using the money, which leads us to back to the maxim of success following success. Funders always want to know about publications and other indicators of research production.

Heilmeier Catechism

The Defense Advanced Research Projects Agency (DARPA) has adopted several questions that its program managers use to evaluate pitches. The Heilmeier Catechism is named after its creator – George Heilmeier – who was the agency's director from 1975 to 1977. (https://www.darpa.mil/work-with-us/heilmeier-catechism).

It is literally impossible to win a DARPA award without putting your proposed project within the framework below.

1. What are you trying to do? Articulate your objectives using absolutely no jargon.
2. How is it done today, and what are the limits of current practice?
3. What is new in your approach and why do you think it will be successful?
4. Who cares? If you are successful, what difference will it make?
5. What are the risks?
6. How much will it cost?
7. How long will it take?
8. What are the mid-term and final "exams" to check for success?

I've encouraged my graduate students and postdocs to frame their research using the Heilmeier Catechism when discussing it with anyone. The first question is especially important (and sometimes hard) to do: describe your science without jargon.

Several prior chapters, especially those on mentorship and authorship, address how to be successful scientists. It is obvious that good mentorship, which leads to happy students and postdocs, superior research, is one of the most important drivers of successful science. In addition, maintaining a good reputation as a fair-player in authorship and data acquisition and sharing shows up when proposals are peer-reviewed. No one wants to fund a louse. Best practices can make the difference between winning and losing in the grants world, and scientists should never underestimate the power of good will. I can't overstate how very thin the line is that divides between winners and losers in grant competitions. Many people, scientists included, don't like to think of research as a competition, but the fact is that a small subset in any grant competition wins the money and the others lose. Meritocracy rules, not egalitarianism, when it comes to funders. Funders nearly always have a choice – they are in the position of power – especially when they are handing out lots money, whether they want to work with a fair player that they like and respect vs. a scientist with a less than stellar reputation. The choice is clear: the best science practiced by ethical scientists has the best chance for winning funding.

Fair play and collaboration

Single investigator grants are not the only choice for funding these days. A higher than ever premium is placed on scientists who play well with others to produce multidisciplinary research with high potential impact. Building teams has become more important than working as the lone hypercompetitive scientist who must control every aspect of research. Thus, it appears that there is somewhat of a dichotomy set up in the psyche of the scientist. On one hand, the team player is rewarded. On the other hand, the competitor is rewarded. I hate to lose, whether it is table tennis or in grants. But at the same time, there is great value in team-building, collaboration, and cooperation. It produces better science and better scientists who can build upon successful teams. Many of my current grants, and the majority of my current budget, come from really productive scientific collaborations. When I'm the PI, my goal is to find the best people as collaborators. But being the PI is a lot of work! Other than additional opportunities for funding, there is another advantage of being a co-PI and not the PI of a grant. The PI has to expend more time and effort to shepherd a proposal through all the channels, coordinate among team members, and generally hold things together: grant administration. Being a co-PI leaves more time for science: maximizing the money to effort ratio. The co-PI might not have as much control over a project, but teaming opens up avenues of science that were not accessible before and often infuses new excitement to a lab. No matter if you find yourself as the PI or co-PI, there are rules of the road that require heeding.

Judge yourself

✓ Do you play well with others or are you more of a loner?
✓ Do you enjoy competition? A fair competition?
✓ Are you a good follower? A good leader?
✓ What is more important: getting money at all costs or heeding your moral compass?

When you are the lead PI

If you want to put together a multidisciplinary proposal and need other people to complete the team and the science, then they must be recruited from the inside or outside of your home institution. If you are looking for the absolute best science, often the person comes from outside your university since no university is great in all sub-disciplines. If a person is going to join your team, he or she must be convinced of at least two things. First, that you will treat collaborators fairly, not only in acquiring the funding but also in doing the project and publishing the results, and second, that they will have a high probability of winning the funds if they enter into the collaboration with you. All your potential collaborators must be convinced it is worth their time to join your team. The most important component in building trust is maintaining confidence and propriety. If you follow the Golden Rule and treat others how you wish to be treated when you are the collaborator or co-PI, then you cannot go wrong.

The recipe for fair play is straightforward. First, it is important to communicate that you, as the lead PI, hold the collaborators and co-PIs in high esteem and truly need them. You must convey that without them the proposal will not likely be funded. If you don't believe that, then there is no need to invite. Second, it is important to communicate that you value their input and time, and that you want to work together as a team. I've made the mistake, albeit rare, of treating collaborators as subordinates. That is the recipe for failure. Third, that as valued team members, you will treat any privileged information and data as such and not share it without their permission. The ethics of authorship come into play in future papers together, and that concept should be communicated. It is not permissible, nor ethical, to obtain collaborators' trust, ideas, preliminary data, and then not include them as a co-PIs and as subsequent co-authors. It is certainly not allowable to cut collaborators out of proposal without their permission. If someone gives data or text for inclusion in a proposal and then decides not to be a part of the proposal prior to submission, it is research misconduct to retain and use the collaborator's data. It is also not allowable to promise funding and then renege from that commitment. If the awarded funding is less than the proposed budget, then of course budgets have to be trimmed. But decreased funding should not be used as an excuse to jettison a co-PI. It is not allowable to mine ideas and data from students, postdocs, and others who work for you without them deriving some benefit from the funding. There are a lot of ethical potholes in funding and seeking funding. Creativity in science is important and people's ideas should be valued. Even after your trainees leave your lab and move onto other places, they deserve authorship on papers as co-conceivers of the proposed science, and their role in experiments including data collection and analyses. They deserve proper credit. Gaining the reputation as an untrustworthy leader, tyrant, thief, or worse is not the route to winning multidisciplinary grants and becoming a trusted collaborator. Sure, bad PIs often remain in science, but they will always land shy of their potential. They are typically not recruited for better positions because of their ill reputations. Finally, the lead PI also has to make sure people feel needed and included after the award is granted. I tell my team that getting the money is relatively easy. Doing what we promised the agency is the harder part. Since funders like to reward high performers, success begets success.

It is not ok for a full professor to pressure an assistant professor for ideas that the former can take and lead as PI. The full professor should encourage junior faculty members to be independent as PIs in their own right. In general, power needs to be used with a good conscience. Administrators should not allow this kind of coercion and bullying to take place within their units. Of course, it sometimes makes sense for a senior person to be the lead PI under certain circumstances, such as when a proposed project is closely aligned with the PI's expertise and for big projects that require an experienced manager/leader. But the decision on who should be the PI should be by agreement and not by coercion.

When you are the collaborator or co-PI

So someone values your input and science – fantastic. And they are doing the hard work of managing the team and submitting the proposal. All you have to do is what you are asked to do for the proposal and the funded project. The best collaborators are responsive and do what they say they will do on time. They understand that without them the project might not have as good a chance of gaining funding. They are trusting and trustworthy to give the lead PI their best data and ideas. They don't compete against their PI with another proposal (though sometimes this happens, but full disclosure is required between teammates). A good collaborator does not wait until the last minute to send documents and needed material to the PI. A good collaborator is responsive to reporting requirements, and manuscript production after the funding is awarded. If you want to continue to be in demand as a collaborator, you must perform up to expectations and be a good team player.

Judge yourself

✓ How trusting are you? Are you trustworthy?
✓ Are you responsive and responsible?
✓ Can you work under deadlines and pressure they create? Do you hit deadlines or do you make excuses why you missed a deadline?
✓ Are you organized?

Recordkeeping and fiscal responsibility

Granting agencies as well as companies award grants and contracts so that the recipient can perform the research outlined in a proposal. In nearly all cases, the awardee is constrained to spend the money in defined cost categories and in a specified timeframe to meet the goals of the stated proposal. To spend funds on other research or in ways not specified in the proposal poses ethical and legal problems (Couzin 2006a). Spending categories can be altered such that a change better fulfills research goals and all parties know and are in agreement. The people that care the most about these kinds of things are auditors. Governments and universities, it seems, have no shortage of auditors. As a rule, scientists probably don't pay as close attention to money issues as they should. Money people at their universities wish, in general, that researchers would be more careful – because of the auditors – but also because good accounting principles are tied with ethics. Since research often moves in directions that are unforeseen, it is arguable that spending funds on unforeseen items is often the most appropriate research decision. Indeed, perhaps the best decision of all is to ensure transparency about funding changes. The PI should not be perceived as being fraudulent when the reality was that scientific items took precedence over fiscal items in the grand scheme of things. Proactive communication is key.

A beginning faculty member or scientist could have a steep learning curve with regards to complicated fiscal and reporting rules. It is important to take the time to learn the rules. Ask questions of administrators, accountants, and faculty colleagues. It is then important to play by the rules. Researchers might be tempted to take the expedient or easy path and then ask forgiveness later rather if caught doing something wrong rather than ask permission in certain cases. As the PI gets older with more at stake if ethical or legal breaches occur, then there may be is a tendency to tighten up practices. No matter what the career stage, there is tremendous value in having the trust of your accountants and administrators; that good will also increases research productivity. Administrators quickly learn which PIs are eager to bend the rules and take shortcuts. Playing it straight has tangible benefits and leads to a less stressful career.

Pushing the limits on proposals

Back to the grant proposal – the people who have the best ideas and the best grant proposal usually wins. When funding rates for any particular panel typically range between 10% and 20%, the competition is fierce. Therefore, PIs could foreseeably be rewarded for either stealing ideas or fabricating data for preliminary data sections (or both!) when composing proposals. The line between right and wrong seems to get blurred in PI's eyes as competition (and desperation) increases. In the whistleblower chapter, we saw that situation in action in the Wisconsin case, and in the case of my own laboratory where "prophetic" data were conceived and offered in the draft proposal as already being full reality. I think many investigators try to forecast data into the future – to state data that will exist at the time when their grant proposals are actually reviewed. While it might be rationalized as reasonable, given all the competition, this practice is actually data fabrication. In reality, I think it is actually a more effective strategy to send an update to the panel manager including any late data prior to the date a panel meets to review proposals. This strategy shows that progress is being made, which can help a proposal. This is a more honest strategy than projecting yet non-existent data into the future.

Stealing ideas is another way that is viewed as a strategy to get ahead. In addition to obvious ethical concerns, stealing ideas from another researcher is also a bad move for one's research career. It is inevitable that the victim of idea theft will someday review a proposal or paper (or both!) of the thief. And, recognizing the thievery, the victim might not be inclined to play fair either and see such as an opportunity for retribution. This is certainly not the kind of research world we want where cascades of unethical behaviors build upon each other.

It is not theft if ideas are obtained in a legitimate fashion, say from a published paper, which can be cited, or hearing a talk, which may or may not be citable. In such cases, it might be more productive to collaborate rather than co-opt the idea outright to maintain good will and build a strong team. Unethical means of gaining a competitive advantage include co-opting data or ideas when acting as a reviewer of a paper or another proposal (see next chapter), or in any case when a communication is confidential and privileged. The most productive researchers I know don't worry much about theft of their own research, but many of them eagerly share stories of being the victims of idea theft. It is a good career move to play it straight with regards to proposals – cooperate and collaborate while maintaining pure competition is not easy, but it does pay off in long and short runs.

Case study 1: the case of a questionable grant proposal (co-authored by Hong S. Moon)

Dr. Jackie Kohls is an associate professor in a microbiology department at an East Coast research university. He is the PI of microbial genetics lab composed of two graduate students, a full-time technician and a postdoc. He was tenured and promoted from assistant professor a year ago; a big reason was because of his several publications and obtaining an NSF grant as a PI and an Army grant as a co-PI. He does not like writing grant proposals, however. He does enjoy the process of extensive literature search and synthesizing new innovative ideas, but he dislikes the tedium of putting together proposals. Therefore, he asks his postdoc, who is being paid full-time from the NSF grant, to write a proposal centered around Dr. Kohls' latest idea; this proposal is to be submitted to the NIH. He tells the postdoc that Dr. Kohls will be PI and the postdoc can be the co-PI. He explains that this arrangement is the only one that would be acceptable to the university since it does not allow postdocs to submit proposals on their own. Dr. Kohls promises the postdoc that continued employment will be made more feasible if the grant is funded. In addition, he argues, the postdoc needs to learn how to write good proposals.

One day, Dr. Kohls was asked to review a manuscript that has been submitted to a microbiology journal. He reviews the manuscript thoroughly and recommends to "reject without further consideration" to the journal editor. He thinks that the manuscript is poorly written and lacks supporting data. However, he finds a nugget in the proposal that supports and expands on his own idea. He forwards the manuscript to the postdoc, along with a literature review that Dr. Kohls has recently completed. Dr. Kohls' idea focuses on the possibility of making a broad-range flu virus vaccine using a specific protein receptor. What was missing from his original idea was a specific activator – which he gleaned from the paper he reviewed.

Dr. Kohls justifies his actions of co-opting the idea by the fact that he needs another grant to keep his graduate students, lab technician, and especially his postdoc, funded and employed since his lab is due to run out of money in less than a year.

He thinks that this concept and approach may eventually result in millions of dollars if it is commercialized in the vaccine industry. He thinks that it would not be stealing someone's idea if he uses different model organism or changes the experimental methods and approaches. Until this point the postdoc has agreed with Dr. Kohls' approach and plan to their teaming to write the proposal. But he objects to the idea of stealing someone else's idea and begins to look for another position. Dr. Kohls is furious and tells the postdoc that there is no need to request letters of recommendation from him – that loyalty and trainee–mentor trust has been breached. After that, Dr. Kohls proceeds to write the grant proposal himself, which is indeed subsequently funded.

1. Funding is one of the most important elements in scientific research. Writing grant proposals is one of the major tasks for most PIs. Does Dr. Kohls do anything wrong with asking his postdoc to write most of a grant proposal but with Dr. Kohls as PI? Is it a misuse of NSF funds to ask the postdoc to write the follow-up grant?
2. Grant proposals are not peer-review publications. Is it ok to be less restrictive for plagiarism and intellectual property issue in writing grant proposals compared to peer-review publications?
3. You have access to someone's idea by reviewing a submitted manuscript. You review the manuscript and reject it. However, you are interested in the concept or experimental designs in the rejected manuscript. The rejected manuscript has never been published later in any journals, thus you can't cite a reference. What is the best way to utilize the idea or experimental designs in your grant proposal?
4. Was it wrong to give the postdoc the rejected journal submission?
5. Was it wrong to penalize the postdoc for conscientiously objecting to participate in the proposal writing and seeking another position?

Case study 2: the case of the collaboration that couldn't

Dr. Billy Block meets Dr. Tommy Tackle at a conference. They've known about each others' work for many years, having reviewed each proposals and papers. They have great mutual admiration; that is not surprising since each professor essentially has very closely aligned research interests and both are productive. Billy Block has the inside track for funding from a large governmental agency center that consists of researchers throughout the country. The center based in Billy's institution, the University of Buttinghead. After a few drinks after the talks, Billy explains much of UB's research proposal content and Tommy Tackle declares that he would enjoy being a part of the center if Billy could find something meaningful for Tommy's lab. When Billy informs Tommy about the rumor that Tommy's own university, the RAM Institute of Technology, is putting together a similar center, Tommy is surprised since no one from RAM IT had contacted him about the opportunity.

After a few weeks, Tommy ponders their enjoyable conversation over drinks. He also learned that RAM IT is not pursuing a center proposal after all. However, the same agency issues a request for proposals (RFP) for mini-centers. Tommy figures that between himself and Billy, the two of them would be a formidable team and would wrap up all the pertinent technologies in the area. Billy thinks this is a good idea, and even though he had not heard about the mini-center RFP, he asks if he can be the PI of the proposal (Block and Tackle thinks he). In addition, he thinks that the UB's center (which is sure to be awarded and which he believes he should have been named main PI) and the mini-center could have great synergy based upon similar approaches. Since it was Tommy's idea, Tommy wants to be the PI and proposes Tackle and Block as a team. After much haggling they agree to be Tackle (PI) and Block (co-PI). After Block's center director hears about this arrangement, he informs Billy that it might not be a good idea to join up with Tackle's mini-center. It could jeopardize their own future funding and set up a conflict of interest. Even though most of Tommy's mini-center proposal is already written with Billy's preliminary data being centrally integrated, three days before the proposal is to be submitted Billy pulls out of the ill-conceived partnership. Tommy Tackle is livid and submits the proposal anyway – with Billy's data included.

1. What are the ramifications of including a potential co-PI's data after the collaboration is dissolved?
2. Why did Billy Block decide to enter and then resign from the collaboration? Was he at fault? What drove his decisions?
3. Do you think there could be an actual conflict of interest if Billy were to be part of his own institution's (UB) center and RAM IT's mini-center?

Summary

✓ Competing for funding requires clear scientific insight coupled with ethical behavior.
✓ Treating collaborators and competitors fairly has tangible rewards.
✓ Fiscal post-award ethics is at least as important as writing honest grant proposals.

Chapter 10

Peer Review and the Ethics of Privileged Information

ABOUT THIS CHAPTER

- Peer review is a long-established system of screening and improving research manuscripts and grant proposals.
- Peer review is based upon trust – that privileged information will not be unduly spread and used to create unfair advantage.

Peer review is among the most dreaded and perhaps the worst thing in science – only surpassed when there is no peer review. Indeed, some people believe that peer review is broken and needs to be fixed (Akst 2010). For now, mainly, it's all we got. For grant proposals and publications, acceptance of a paper or project usually relies extensively on the responses of other scientists who are specialists within a particular field of study. Science is indeed a grand community in which its members are responsible for quality assurance in publishing and allocating grant funds. Oftentimes, the same day that I submit a paper for publication, I'm recommending either the publication or rejection of someone else's manuscript that has been submitted to the same or another journal. On a day such as that, scientists appreciate the smallness of the professional world that we inhabit. As a journal editor, I doubly appreciate it. I see the manuscripts authors write and the peer reviews offered by other scientists and associate editors. It is a system I'm happy to participate in. While altruism and the Golden

Research Ethics for Scientists: A Companion for Students, Second Edition. C. Neal Stewart, Jr.
© 2023 John Wiley & Sons Ltd. Published 2023 by John Wiley & Sons Ltd.
Companion website: www.wiley.com/go/stewart/researchethics2

Rule are often the norm, nearly every seasoned scientist has a story of how ideas and projects were "stolen" as the result of peer review; clearly, unethical abuses of privileged information happen. Also, stories abound about unfair or even cruel treatment by reviewers, editors, and funding agencies. Therefore, it is important to delve into the rules and best practices of peer review. Examining pertinent case studies will help to unveil of mystery that is peer review. There will be an emphasis on the nature of privileged information. There are several current discussions and "experiments" in the worlds of granting and publishing to reform peer review, and these will also be examined.

The history of peer review

Prior to the formation of the Royal Academy of Science in the 1660s, the medical and scientific arts existed as the work of very few people, a relative microscopic entity compared with today's gigantic world-wide science infrastructure. In those days, scientists were very loosely organized with some evidence that forms of peer review were used to evaluate the quality of research (Spier 2002). The first scientific journal and subsequent peer-review system came into being with the *Philosophical Transactions of the Royal Academy of Science* in 1665. But as Spier (2002) states, until the middle of the twentieth century, there was more potential journal space than there were submitted articles, and the breadth and depth of science was relatively narrow. Journal editors, or course, read the articles that were submitted, and they could also circulate submitted papers to a number of society members. Submissions were typically passed around and informally reviewed. Incidentally, all members likely knew each other well. After World War II as science boomed and geographically expanded beyond old schools and cities, such an informal system was no longer workable. The numbers of scientists, journals, disciplines, and their subdisciplines all increased dramatically; all were driven by increased funding to spur science competitiveness and the reach of science. Generally speaking, as funding increases in an area, productivity; i.e., the number of papers produced, typically also increases with more scientists being trained. Sometime in the mid-twentieth century, there were more papers being generated than space to publish them, a situation that continues today. However, it can be argued that with many journals being published strictly online, there no longer exists any real physical space limitation, but the need to select only qualified articles remains a standard, at least for most journals. Indeed, the main function of peer review is quality assurance.

The nature of journals and the purpose of peer review

There is a virtual stratus of scientific journals. They range from "high-impact" journals that publish a variety of papers that many scientists, irrespective of discipline, would be interested in reading to geographically centered and discipline-narrow specialty journals that fewer people read. There are numerous journals in between these two disparate strata, including very well-respected discipline-specialty journals that are international in scope. Almost all scientists want their work to be read and appreciated; therefore, the higher profile journals are generally more desirable targets for publishing, but they are more difficult to pass editor and peer review. In practice, most scientists publish most of their research in specialty journals. Therefore, it is no surprise that the rejection rate of elite journals is quite high, say, less than 10% for some. Conversely, almost all papers are accepted in many lower-tier regional journals. Therefore, it is not surprising that the nature of peer review and editorial decisions vary widely among journals.

Open-access journals vs. subscription journals

In Chapter 8, open-access journals were introduced in light of predatory publishing (predatory journals are also open-access journals). There are thousands of legitimate journals that are open access. Authors of papers published in open-access journals pay article publication charges that are typically thousands of dollars. In exchange, the authors retain their copyrights and the article becomes available for public viewing immediately. In traditional subscription-based journals, the authors don't pay a fee, but relinquish their copyrights. These traditional journals also typically offer "hybrid" open-access options if authors wish to pay. If not, papers published in these subscription-based journals are behind "paywalls." Since most research is funded by public funds from governments, there is an ethical argument that research from this funding should be made immediately available. One movement to do so is "Plan S" (coalition-s.org), whose membership includes mostly European funders. While the biggest push is publishing in open-access journals, another means to make research immediately available is for authors to post their accepted articles onto public repositories. Indeed, the US White House Office of Science and Technology Policy recently posited that all government-funded research should be publicly available upon publication starting no later than by 2025 (Tollefson and Van Noorden 2022). Herein, the potential solution is public repositories to post peer-reviewed accepted manuscripts. In theory, this solution has some advantages of open access. Indeed, two drawbacks of the open-access journal model is 1) high additional costs to authors and governments, which include equity issues, and 2) since journals only get paid when articles are published, there could be unethical pressures to accept papers that are not meritorious. This second problem can lead to corruption of the publishing process. Has it occurred, beyond the obvious predatory journals that don't perform the normal rigorous activities associated with academic publishing? I think that it is inevitable to some degree, but the drive for journal to increase their impact factors (see below) strongly offsets gratuitous acceptance of bad papers in open access journals. The first issue – the growing exorbitant cost of article publication charges – is perhaps more problematic. For rich countries and governments, funding agencies and universities can largely afford these, but what about poorer countries? Unless the journals waive these charges (and some do), it may cost more to publish an article than an African scientist makes in a year. Posting papers on a public website is practically free.

Bibliometrics and the rise of journal impact factors

Journals are largely stratified within disciplines according to the journal impact factor and other similar indices based on citations over time. A journal's impact factor for any particular year is the number of times papers published in the previous two years were cited that year divided by the "citable items," which equates roughly to the number of papers published over the past two years. Each year, impact factors for the previous year are published by in the Journal

Citation Reports and journal editors and publishers pay attention to progress (happy!) or slippage (sad!!) in impact factors and the relative rank of a journal in a scientific field category, such as "plant science." While we're not supposed to think it (and especially not say it), the "best" journals, those with the highest impact factors, are in highest demand for paper submissions. They are the journals of choice in a discipline. Therefore, editors, who always want the best papers for their own journals, can try to make their journal desirable. The impact factor is one aspect. Other aspects include how fast peer reviews and decisions are made, and if similar papers are published in the journal. Editors can come up with "special issues" wherein papers within a topic are published together, making the set more attractive. Of course, some journals are too new to have impact factors and be indexed in the Web of Science or other compilations. Some journals are not indexed because they might be too obscure (regional) or published in languages other than English, the de facto official language of science. The bottom line is that scientific publishing is competitive (Handwerker 2010).

For most of us scientists, we hope to get our papers in the best journals we can without too much haggling with editors and reviewers. The vast majority of authors want their studies to result in papers that are sound and publishable (published). Some scientists aspire to create consistently great papers that are published in prestigious journals. No scientist wants to be embarrassed. Many scientists say they don't really care about impact factors or that impact factors don't equate to journal or publication quality. Some of that is true. A paper that ends up being relatively obscure and cited little may occasionally be published in *Science* or *Nature*. But most scientists I know still pay attention to impact factors. Scientists want their work read and cited (Handwerker 2010). Only dilettantes in science publish because their boss tells them to or for the purpose of simply gaining tenure.

I think just about everyone who knows about impact factors and cares about them, even if they profess otherwise. The proof is in the eating. Typically authors will submit a paper to the "best" journal (read highest impact factor) that they think will accept it. If they make the mistake of aiming too high and the paper is rejected, they will reformat and send it to lower tier journal and so on. This is a very common practice that wastes a lot of time for authors, editors, and reviewers.

Of course, there are lots of caveats to assessing impact factors. Audience, numbers of good journals competing for good papers, and sexiness of field are among the factors to consider when judging or comparing impact factors. As an exercise, we could, for example, compare publishing in the fields of cancer research and soil science. The top biomedical journals that cancer research is published will have much higher impact factors than the top soil science journals (by over 10-fold). There are also many more cancer researchers than soil scientists to cite papers in their respective fields. These biomedical scientists are supported by more money than soil scientists, therefore more science per scientist is produced in the biomedical field compared with soil science. There are many more factors that could be discussed, but the important point is that within fields, there is variable supply and demand among journals. This leads to different cultures among fields and journals. And, therefore, editors and reviewers have various goals for reviewing papers that go beyond an assessment and assurance of quality. Many people consider the rejection rate of a journal to be an important factor of stringency and quality. Increasingly, journal editors select papers to published based on novelty and the size of the study. They also prefer striking and impactful studies that will make a splash in the field.

Goals of editors and reviewers

Editors of elite journals don't even send most of the papers they receive out for review. They merely reject it after a quick read since they know that most papers don't fit what they are looking for. *Science, Nature, Cell,* the *Proceedings of the National Academy of Sciences USA* (*PNAS*) are among the journals that reject a lot of papers without review. So, when a paper is sent out for review from one of these top journals, the reviewer is aware of certain salient facts. I review between 10 and 20 papers per year and very few are submitted to *PNAS, Nature, Science,* and other top-tier journals. Here were the things I looked for and thought about while I was performing the review. The main thing was to determine whether the paper is "good enough" for *PNAS*. The word "good" used here means various things. Is the study going to be of interest to a lot of readers – scientists in a broad range of fields – or just those within a subdiscipline? Is the paper written so these various scientists can understand it? Is the paper a significant "breakthrough" or is more of a "standard" incremental advance? Finally, and really most importantly, are the data sound to support the main points the authors make. Peer review doesn't guarantee a paper's findings are true (Wilson 2002). No matter what the journal, the reviewer's main job is to assure the data presented are sound, and of course, this assumes that the authors are being honest (Wilson 2002).

As an example of how peer review works, for the *PNAS* (impact factor = 12.78 for 2021) submission I recently reviewed, the topic was in a very exciting field, the manuscript claimed a very big result that would get noticed by a lot of people, but the data were not sufficient to justify the grand claims the authors made in the submission. As a peer reviewer, I recommended rejection and explained all the reasons that led me to that recommendation. It is very important in big papers to make sure that there are no big holes. If this paper were to be published in *PNAS*, the lead author's university would likely issue a press release citing that their important study was "published in the prestigious journal *PNAS*" or similar language. Science writers and non-scientific journalists would use such press releases to write popular articles about the new data and findings. Aunt Betty will read about it in her local newspaper and ask me questions about the study. Potentially, a company might license the technology, secure funding and try to make and sell products from the findings. They will use the published study as leverage and evidence that their product is genuine, useful and worth the asking price. Potentially, an environmental group will cite the study as proof that a certain technology will harm the planet. Legislators will be swayed about making new laws to protect the public. All these trickle-down effects start with the science and its ensuing publication. Therefore, bigger papers will have bigger real impacts and consequences. The truth is, however, most scientists publish very few high impact papers.

No matter what the apparent status of the journal, editors and reviewers should (and usually do) have very high standards for papers they decide to publish. They still sometimes make mistakes, but they rely on stringent reviewers and reviews to maintain their journals' reputations. Back to our example. Having been rejected, the authors of the rejected paper will do one of two things. Their first option is to reformat the manuscript and send it to a more specialized journal, maybe one that has an impact factor of 5 (still good) or less, but the paper might not receive the recognition of the higher-impact journal.

But this particular paper might still get rejected (see below). Their second option is to perform more experiments and provide a fuller picture of the result they initially claimed and resubmit to *PNAS*. This might or might not be reasonable or even possible. They might have already shown their very best data and likely the best data they can hope to collect. Sometimes things don't work out in science the way scientists hope.

This week I also received and reviewed a paper from a very good, but more specialized journal (impact factor = 3.5). My goals and stringency for reviewing this submission were decidedly different from the *PNAS* submission. In addition, I also shared the manuscript with a postdoc in my lab, asking him to also perform a review on it for training purposes. It also was relevant to his current research. I asked him to keep the paper confidential. For the paper submitted to the specialized journal, I'll ask fewer high-level questions. Once published, chances are the findings appearing in the "lower journal" likely won't be repeated in the *New York Times* and many fewer scientists and other people not in the specific field will take notice. It will be read and perhaps cited up to 20 or 30 times in the next 10 years or so, maybe more, maybe less. It will be an incremental increase in knowledge. I read the submission and suggested ways the authors might improve it. I asked for certain missing details, pointed out some typos, indicated how a badly labeled figure probably needs to be re-rendered and I also questioned a statistical method used – I thought that another method would be more appropriate and useful. I gave my reasons. It was obvious that this particular submission was a significant part of a graduate student's degree research and she and the other authors will be happy when they hear from the editor that I (along with my postdoc) and another reviewer or two recommended that the paper be published pending minor revisions. Submissions get rejected about half the time from this particular journal. Some of these rejected manuscripts are totally revamped and resubmitted to the same journal, and some of the rejects are reformatted, probably altered to address reviewers' comments, and then submitted to another journal farther down the impact factor food chain.

All the above discussion about journal choice and author behavior would seem to indicate that journal impact factor is somehow tied to the quality of science or the impact of particular papers or scientists. Most all commentators about bibliographic metrics agree that this is not the case (Van Noorden 2010). Impact factor simply indicate how many times the average paper is cited in a particular journal, and says nothing about any particular paper.

When journals don't publish

The song and dance has been the same for a long time. Step 1: Author submits a manuscript for publication. Step 2: Journal editor sends the manuscript out for peer review. Step 3: Paper is accepted and published in the journal. Or alternatively (step 3) paper is rejected and the authors start over. Step 4: If the paper is accepted, and if the journal is an open access journal, authors pay the publication fee. Now there appears to be another song and dance adopted by the journal *eLife*, which was founded in 2012 and has an impact factor of 8.713 as of 2021. Step 1: Authors submit a manuscript for peer review to *eLife* and pay $2000. Step 2: Manuscript is peer reviewed. Step 3: Authors may or may not respond to the peer reviews in another version. Step 4: *eLife* neither accepts nor rejects the paper but rather posts it online along with the peer reviews (Brainard 2022). In adopting this new model, the journal editors forgo what many authors (and the scientific community) expect from a journal: namely for an expert decision on whether to accept or reject a paper based on quality. Furthermore, also abdicated is the associated canonical status of publication. The "paper" would now appear to be the same as a manuscript posted on a preprint server, but now accompanied with peer reviews. In my view, I would never have an interest in authoring such a paper or even citing a paper that has undergone such a process. I also have no interest in reading peer reviews. Perhaps this model will be adopted by other journals. But I hope not. I'm quite satisfied with the status quo.

Which papers to review?

Editors pick and request reviews from scientists in the field. Perhaps it is human nature. I always accept the chance to review submitted papers in high-impact journals, and most often review for medium impact journals and I don't typically review papers from really low impact journals (seldom for below an impact factor of 1.0). I review the papers that are most interesting to me or because I want to help the editor. Some exceptions to my "rule" are reviewing for new journals that are in an up-and-coming area and not issued an impact factor yet, or articles that are particularly interesting to me. I figure that I should review about twice the number of papers per year that I publish. I rationalize that each of my manuscripts gets sent out for at least two reviews, hence reviewing twice as many papers as I publish would be minimal service to the profession. If I were to really be fair to the scientific community, the number of papers I review would be a good bit higher, since some submitted papers get rejected and must resubmitted and re-reviewed. I spend probably an hour or two reading and reviewing a paper – sometimes longer for more complicated papers. It takes significant commitment of time and effort to perform fair reviews. No matter what the journal, their papers will never be better than the quality of the peer reviews, and so peer review is a service that scientists should take seriously.

Open reviews and discussion

Some journals are experimenting with non-traditional peer review. In traditional peer review, the authors are known to reviewers but peer reviewers are anonymous. Some journals now make the names of authors blinded. Some journals publish the names of reviewers if a paper is accepted. Of course, peer reviewers can always give permission to waive anonymity, but most don't for various reasons including the fear of future retribution. It seems to me that providing non-anonymous reviews is a way to maintain higher civility and forthrightness in the review, and a path for greater accountability. For some reason, reviews can get nasty. Perhaps that is one driver behind all the nontraditional peer-review models. But then again, traditional peer review works more often than it doesn't. In addition, as an editor, I often have a difficult time recruiting peer reviewers to review a manuscript. I can't imagine adding in complications that make this peer reviewer recruitment even more difficult. In addition, it is also up to editors to assure that peer reviewers aren't nasty in their comments. For those reasons, I prefer the traditional peer review model. I think most scientists are happy with the current peer-review system used by most journals and accept that, while flawed, it is better than any alternative we have yet conceived. But of course, people are innovative in gaming the system (see the box below for one example).

Performing your own peer review

Given that journal editors do have a difficult time finding peer reviewers, most journals will ask the submitting author to recommend potential peer reviewers. As long as the authors don't suggest people with a conflict of interest (friends, family, coworkers, and collaborators, to name a few), then the practice is helpful. Some journal editors have a policy that no more than one author-suggested peer reviewer provide a review. A few years ago some "innovative" authors came up with the idea of suggesting themselves as peer reviewers. Of course, they didn't name their own names, but a fake name connected to a real email that belonged to the author. "This is the best study I've ever read" is what I imagine the fake peer reviewer wrote. "It must be accepted for publication." Journal editors eventually caught-on that this was happening. One defensive measure they took was to check email addresses as well as accepting only institutional email address domains and not private email addresses, say, with yahoo.com or gmail.com domains. Journals that use catalogs of verified email addresses also helps stem the problem.

Judge yourself

✓ Do you enjoy critiquing papers and experiments? Are you able to offer constructive criticism without being mean?
✓ How do you feel about journal impact factors and research metrics in general?
✓ How personal do you take criticism? Can you separate your work or product from ego and self-worth?

Grant proposals

Grant proposals get peer-reviewed too. There are two big differences between granting agencies and journals. For granting agencies, they have a finite amount of money and wish to distribute it to the PIs who proposed the best research. For most journals, a manuscript must simply meet certain criteria to be accepted for publication. Yes, there may be finite journal space, but journals don't have the hard constraints of granting agencies with a defined pot of money to spend. Therefore, grant proposal reviewers are really looking for reasons to reject a typical proposal, say 80% or more of them. Therefore, all parts of proposals are scrutinized, including researchers' CVs, statements of facilities and equipment, and current and pending funding in addition to the proposal narrative. The second difference is that a grant panel usually wishes to reach consensus on which proposals get funded. A journal editor simply has to make an executive decision about publication and what changes need to be made to a paper.

Confidentiality and privileged information

All peer-reviewed information is considered to be privileged, confidential, and not to be exploited by the reviewer. Therefore, authors' names, ideas, data, materials, and technologies divulged in a paper or grant proposal should be handled with the utmost care. Information provided in papers cannot be used or acted on until the paper is published. Information in grant proposals should always be held in confidence in perpetuity. Note that I do request oftentimes that my trainees assist me in providing reviews. But I request that they not further distribute materials or unethically utilize the divulged information. If there is any question about propriety, we have a discussion to clarify the ethical boundaries. As noted earlier, many scientists feel that confidence has been betrayed on occasion by suspicious timing of publication of coincidently similar work ("they stole my research!"). Then again, I've observed very similar, yet equally superb papers published simultaneously in the same journal where the authors of the two groups are on opposite sides of the world. Sometimes a particular field as a whole moves in concert as great minds think along the same lines. This seems to happen with regular frequency in science where there is a complete absence of unethical behavior on anyone's part. I'm not minimizing the importance of maintaining confidence – peer review becomes ineffective when confidence is broken or when data or key concepts are leaked. Scientists get ideas from all kinds of sources – published papers, meetings, and everyday occurrences. Clearly, the line must be clearly drawn so it is not crossed. Researchers must clearly delineate ethical guidelines before the opportunity to cheat presents itself. The Golden Rule is a good guide to keep in mind. If someone co-opted your particular idea or data that was submitted in a grant proposal or manuscript, how would you feel? If it is too close to that presented in your paper, you might feel violated or exploited. If it is the next logical step and a significant diversion or application from your own, you might not be offended at all. You might even feel flattered that something you wrote spurred on work that you would have never thought of. However, to play it safe, reviewers should never retain submitted papers as resources or any notes on the papers. Review and discard is a good rule to follow. In addition, reviewers do well to have a short memory or at least wait until their reviewed paper is published before they act on any urges to use the information provided in confidence.

Reviewers

It is almost never appropriate for a co-worker, collaborator, co-author of a recent grant proposal or paper, mentor, or mentee to serve as a peer reviewer. This is a conflict of interest (see Chapter 12). Of course, there are always exceptions to some of these situations, and the potential reviewer should always make sure there is full disclosure to the person inviting the review. The point of peer review is to obtain an objective and non-biased recommendation based on quality. Someone with a conflict of interest oftentimes cannot overcome an inherent bias. If I receive a paper from an editor and, say, one or more of the authors have been co-authors of mine in the past, and I still want to review the paper, and I feel like I can be objective, I'll disclose these facts to the editor/program manager and let him or her make a decision on how best to proceed. Authors of papers and proposals often have the opportunity to recommend reviewers. It is poor form to recommend your friends and collaborators as reviewers. In instances when I'm serving as an editor, and I discover this has happened, I have diminished opinions about the professionalism of the communicating author, since I feel that confidence is attempted to be betrayed and objectivity is about to be subverted. I'm not a fan of being taken in a shell game.

Judge yourself

✓ Can you keep a secret? Are you trustworthy with confidential information?

✓ Do you have a short memory? Can you easily recall concepts and their origins?

✓ Do you feel desperate for ideas or do you feel as if you have more scientific ideas than you can possibly pursue?

Final thoughts

Peer-review works best when it is played straight up with no hidden agendas – from authors, editors, or peer-reviewers. There is no place for cronyism, idea theft, malfeasance, or dirty dealing in science. As authors we should value strict and stringent reviewing and meritorious decisions. As reviewers, we should value the best science and try to influence authors and editors to publish the best papers possible. I serve as author, editor, and peer reviewer. Each role is vital to ensure sound science.

Case study: what is responsible peer review?

courtesy of Ruth L. Fischbach, PhD, MPE and Trustees of Columbia University in the City of New York; http://www.ccnmtl.columbia.edu/projects/rcr

This case was adapted by Columbia with permission from: "Reviewer Confidentiality vs. Mentor Responsibilities: A Conflict of Interest" Research Ethics: Cases and Commentaries, Volume 3, Section 3 Brian Schrag, ed. Association for Practical and Professional Ethics.

Dr. John Leonard is one of very few molecular biologists working in a particular field. Dr. Leonard receives a paper to review, about a protein called survivin, which he and a graduate student in his laboratory are researching. The article was submitted by Dr. Mark Morris to *Protein Interactions*, a medium-impact journal, and the editor asked Dr. Leonard and two other experts in the field to review the paper. The article suggests a new interaction between survivin and the protein GFX and provides evidence for the fact that both proteins are necessary for the full survival-promoting function of survivin in a cell. The article also describes, though, that if there is too much survivin inside cells they die.

But the paper is fraught with problems: poor controls, inconsistent data in figures, and alternative explanations are not considered and claims are overstated. Dr. Leonard gives the paper to his graduate student Melissa Zane, who gives it a detailed critique and recommends significant revisions. Ms. Zane has never reviewed an article before, and Dr. Leonard thinks that doing so would be a good educational experience for her. Ms. Zane notes the finding about too much survivin being toxic to cells, a problem she has had working with the protein, and discusses it with Dr. Leonard. Both agree that they should lower the dosage of survivin in her experiments; the cells actually survive for a week, longer than her experience before, and then they die.

Dr. Leonard submits Ms. Zane's and his own comments about the research to the editor, suggesting that the paper be accepted only after a few more experiments are performed to validate some of the conclusions. One of the other reviewers has comments similar to Dr. Leonard's, and the editor asks Dr. Morris, the author, to make the revisions before he will accept the paper.

But in the next few weeks the interaction between GFX and survivin that is discussed in the paper remains in Dr. Leonard's mind. GFX was not a line of inquiry that Dr. Leonard and Ms. Zane were following in their research. They were focusing on other stimulatory proteins, but unsuccessfully. Dr. Leonard suggests to Ms. Zane that she add a compound to the cell culture system that stimulates the cell to produce its own GFX, a method that is somewhat different from what was in the paper by Dr. Morris that is under review. The enhancement method works. The cells live for a month.

Ms. Zane and Dr. Leonard draft a paper based on the results, which includes appropriate controls. *Science*, a prestigious journal, accepts the paper. Several months later, *Protein Interactions* publishes a revised paper from the laboratory of Dr. Morris. But after Dr. Morris sees the article in *Science* he suspects that Dr. Leonard, who was an anonymous peer reviewer on the paper, might have taken some of the ideas for the *Science* article from his paper under review. Dr. Morris knows that Dr. Leonard had not been working on GFX because it was hard to purify and deduces that he used material in the unpublished manuscript to stimulate GFX activity.

1. What types of conflict of interest might arise when someone is asked to review a paper or grant application?
2. Is it ever appropriate for a peer reviewer to give a paper to a graduate student for review? If so, how should the reviewer do so?
3. Is it appropriate for a peer reviewer to use ideas from an article under review to stop unfruitful research in the reviewer's laboratory?
4. Is it ever appropriate for a reviewer to use ideas from a paper under review, even if the reviewer's method to achieve a result is different from that used in the paper under review? If so, how should the reviewer proceed?
5. What are some of the challenges in the current peer-review process, in which the peer reviewer is anonymous, but the author is known to the reviewer?
6. What recourse is there for Dr. Morris if he suspects that his ideas were plagiarized?

Summary

✓ Peer review is a pillar in science.
✓ Responsible peer review is everyone's job.
✓ Maintaining confidence is critical for effective peer review.

Chapter 11

Data and Data Management: The Ethics of Data

ABOUT THIS CHAPTER

- The sustained integrity of data is foundational to the integrity of science.
- Data should be archived in a way to maintain their viability and accessibility.
- Data should be made available for sharing among researchers.
- Digital data present certain challenges, especially when compiled in large amounts.
- Data wikis and other shared vehicles for data deposition are becoming more widely used.
- In many fields, sharing materials through material transfer agreements (MTAs) are as important as sharing data.
- Unpublished data presented at scientific conferences may have some unique challenges.
- Data collection, curation, and analysis must be performed in ethical ways that are transparent to future users.

At the center of science is data. Except for perhaps in some of the most theoretical areas of science, all scientists think intensely about collecting, analyzing, protecting, and then disseminating data from their research. Hopefully, they plan also for its preservation and utilization for the use of other scientists. Data integrity is critical, including analysis and storage. Thus, for this chapter, the focus will

---

Research Ethics for Scientists: A Companion for Students, Second Edition. C. Neal Stewart, Jr.
© 2023 John Wiley & Sons Ltd. Published 2023 by John Wiley & Sons Ltd.
Companion website: www.wiley.com/go/stewart/researchethics2

be on three issues. The first is to examine the landscape of archiving and making data available and usable for the greater research community: data preservation and stewardship. Sharing summary data happens at the point of publication, and the dispersal of raw data occurs almost never in some fields to always upon publication in other fields. A vast continuum exists post-publication. Indeed, this might be *the* big issue in many areas of science now where giant data sets are easily produced. The second big issue is the simultaneous need of scientists to both present data in meetings, yet attempt to retain exclusive rights for publishing. It is interesting to look at the shifting sands of technology and ethics of walking the tightrope of presentation ethics. Third, and perhaps most important, is the nature of data curation and analysis, including appropriate statistical analysis for honest reporting (and not research falsification).

In the minds of most scientists, data collectively represents a sacrosanct and precious commodity – perhaps the most valuable raw resource in science. Grant proposals are written to collect and analyze it, and careers are made on publishing it. Yet, the rules surrounding the handling of data are still, at best, squishy. Despite the importance of data, there are seemingly few resources available to assure its long-term availability and accessibility. The same is true for archival research materials (Couzin-Frankel 2010). I think this is one of the main reasons that spurred the US National Academies to pen a volume on *Ensuring the Integrity, Accessibility, and Stewardship of Research Data in the Digital Age* (National Academy of Sciences 2009). The title says it all. Since data are so important (data, plural; datum, singular – but no one really ever seems to refer to "datum"), it compels us to examine best practices for us as individual scientists and the science community as a whole. We also need to look ahead to how technology will affect how data are stored, presented, and accessed in the coming decades.

Stewardship of data

I recall a manuscript of mine that was rejected. One of the main reasons why it was rejected was that my co-authors and I did not spell out the data deposition plan. Indeed, one reviewer pointed out that we had shown very little of our data in the paper and thus the readers would likely be interested in accessing more (or all) of it. Most funding agencies require data collected with their funding be made public. Most journals and scientists assume that when a paper is published, that data described will be made available to other researchers. How are we doing on this front?

Data of old

In the old days, read, before computers were widely used to collect, analyze, and store data, there was less data and its accessibility and archiving were more ad hoc than today, at least in many fields. Data were collected largely by hand, placed in notebooks or loose-leaf data sheets, and they ended up in physical files stored in file cabinets. Indeed, many scientists still do this. All scientists (students, postdocs, research associates) in my lab, and, I'd guess, people in most labs, are required to keep bound laboratory notebooks written in ink as official documentation of their research. The notebook pages are numbered and the notes and data may be witnessed and signed off periodically by someone else in the lab. These ultimately belong to the institution. Keeping good lab notebooks is vital to scientific integrity and stewardship. Bound notebooks remain foundational for intellectual property protection – to prove that an invention happened when and what inventors may claim (see Carlson 2010). Notebooks are important also as scientific records; in fact, that is their primary importance. So, when people leave my lab, their notebooks stay behind with me and we indeed have physical file cabinets containing many lab notebooks. So, what happens to the notebooks when I move on from my current university, retire, or die? The answer is not clear (Brown and Nguyen 2021). Ideally (in theory), the notebooks would remain with the university where an archivist carefully catalogs the documents. In practice, I doubt this happens much of anywhere. I know of no people at my university to take care of my documents when I die. Most typical, I suspect, is that the PI either takes the most important ones to the new locale or home upon retiring. But I seriously doubt I'll move all those file cabinets full of data home for me to pore over during my retirement. I suspect that laboratory notebooks from a career are most often eventually destroyed when the PI retires or dies. This occurs because (as fond as we are of pontificating about the importance of data stewardship) no one wants to pay for long-term storage and cataloging of laboratory notebooks or for the people necessary to organize the data so that they are perpetually accessible to other scientists.

Scientists are still seen as mom and pop entrepreneurs with every man (and woman) for him(her)self. The same is true for scientific materials – it is difficult to archive items beyond the life of a lab (Couzin-Frankel 2010). The hope is that most of the meaningful raw data in lab notebooks is reduced to meaningful figures and tables and published in peer-reviewed literature, which is readily accessible and citable. Therefore, if crucial methods are captured in publications, and if the raw numbers are averaged and presented in papers as reductionist displays, that data integrity and stewardship is addressed sufficiently in most cases, at least in fields where data are still relatively few and easily managed. Significant amounts of data are still collected and handled this way. Many journals have taken up the practice of allowing authors to include supplemental data online. This is a terrific opportunity to make data publicly available for downloading. There are at least two issues that researchers should pay attention to in this regard. The first is that the raw data are archived. In many cases, it is appropriate to transform data and summarize it in various ways, but raw numerical data should be preserved. The second is that metadata should also recorded to enable the understanding and future analysis of the raw data. Analysis must be replicable by people not involved in the collection of the original data.

Judge yourself

✓ How do you feel about sharing data? Some people are data distributors "Johnny Data-Seed" and some are data hoarders. The latter feel that they may want exclusive use in perpetuity. Do you fit into either one of these molds?

✓ How organized are you? Do you meticulously collect, record, and organize your data so that it can be shared and understood by others? Do you think to include metadata?

Stewardship of digital data

So, there is more to my story of my rejected paper, which illustrates issues surrounding the storage of digital data in the field of genomics. It might help the non-genomicist to briefly review the timeline of DNA sequencing technology. The first DNA to be sequenced was that of a bacteriophage in the early 1970s. Until around 2005, the technology for sequencing genes had undergone a few incremental improvements. While the amount of data grew exponentially, the kinds of data output were essentially static. The original idea from the 1970s was to clone a segment of DNA and sequence it, giving read lengths per run at about 800–1000 bases for single contiguous molecules. Essentially, these experiments could be performed by hand and data read visually on autoradiograms. Later on these data were collected using a more sophisticated automated capillary electrophoresis instrument, which facilitated faster data acquisition, and then analyzed and stored by computer. The DNA sequence data could essentially be pasted into the laboratory notebooks. As more scientists used personal computers, and then when the internet became accessible, the US government (and then other governments) created digital data repositories; e.g., GenBank, which was started at Los Alamos National Lab. Here university, government, and industry scientists (if they wished) could deposit to GenBank the sequences and their functions if known, and the donor organism name. It was expected that if a researcher published a paper on a gene, that its sequence would be deposited into GenBank. All was well. But in the mid-2000s the genomics revolution accelerated as several research and development projects resulted in new instruments and technologies to generate much more sequencing data at a much faster pace (Figure 11.1). This resulted in even more DNA sequencing data. For example, in 2000, the year the first complete draft of the human genome was published, here were 8 billion base pairs of human genomic data deposited in the main databases in the United States, Europe, and Japan. Ten years later there were 270 billion base pairs in the system with the

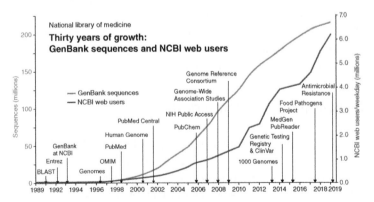

Figure 11.1   The US National Library of Medicine's history of DNA sequence data deposition over time versus number of users.

doubling of data every 18 months (Anonymous 2010). The technologies are so powerful now that small genomes can be sequenced in a day and the entirety of the expressed genes for more complex organisms such as plants and animals could be sequenced in a week or less as cost for sequencing has drastically decreased over the past two decades, which has continued into the 2020s given the advances in next-generation sequencing platforms. Of course, it would take longer than a week to analyze and make sense of the data, but the point is, whereas genes were once being sequenced one at a time, now all of them in an organism can now be sequenced in parallel. The result was now the sequences of millions to billions of bases of DNA can be generated at one time. And this is where my paper had a slight hitch.

My group and now other individual (regular) labs can sequence the transcriptome (expressed genes) or the genome (all the DNA in an organism) for a plant using standard funding by sending the RNA or DNA off for sequencing and then receiving back millions to billions of bases of data. So, for my paper, we assumed when we wrote it that we would deposit the data in GenBank or some other repository for public access, but we never stated that intention in the manuscript. The fact is, at the time we were actually toying with the idea of making our own database that could be expanded and organized to better serve the research community interested in the certain kinds of plants we study. We figured we'd make the decision after the paper was accepted, a big mistake. So, in the resubmitted manuscript, we stated that we would simply deposit the data into GenBank. Biologists assume that GenBank can take all the data, ad infinitum, which is probably true. Perhaps also complicating the review of our paper was that our collaborators and funders for the project were from a company. Maybe the reviewers suspected that we wanted to publish the summary results from the big experiment using the big data set, yet keep the raw sequence data a secret. Would that be allowable? For us, no, it was not allowable given that making the data known to others would be more important than preserving private commercial interests.

Data from published work and publicly funded research must be made public

The primary duty of scientists is to collect, analyze, interpret, and publish meaningful scientific results, most of which is based on data. One key doctrines of science is that science should be able to be replicated. Another is that science demands that "all the cards be placed on the table" when a paper is published. Or as the National Academy of Sciences (2009) succinctly stated, "Researchers are expected to describe their methods and tools to others in sufficient detail that the data can be checked and the results verified." It is difficult, if not impossible to meaningfully criticize a study if data are not shown. Indeed, there are instances in papers where authors cite "data not shown" when making a point, but it is usually relatively trivial instances when it might not be appropriate or within space constraints to show those particular in a paper. This is why supplemental information that is not on the journal's printed pages, but rather deposited on a journal's website is becoming more prevalent. Nonetheless, many funding agencies require that data collected from their projects share data with other researchers. Optimally, that data be placed in a public database (National Academy of Sciences 2009). But this requirement is not universal. From a stewardship perspective, I don't know of good arguments to the contrary except to protect human subject details and for classified research.

In the United States, the NIH has taken the issue of public access of data to another level by requiring that their grantees deposit their manuscripts that are accepted for publication into PubMed Central. While publishers may not appreciate this requirement because anyone may then download the paper for free, it may potentially disrupt many publishers' economic models of business. Nonetheless, publishers are increasingly setting up systems to allow people to read papers behind a paywall even if they don't allow it to be downloaded and saved. But as we saw in Chapter 8, Plan S and other mandates from governments and funders may render paywalls eventually obsolete.

Data must be analyzed appropriately to prevent falsification

If raw data and metadata are managed and archived correctly, then other researchers ought to analyze those data and arrive at the same conclusions that the data depositors did in their paper. That is to say, that statistical analysis should be chosen and performed correctly. There are several issues that can send scientists down the wrong path. One is referred to as HARKing (hypothesis after the results are known). While it is not specific to data, the hypothesis chosen may affect data analysis and statistical methods utilized. If researchers present the analysis as an a priori one, rather than a retrospective one based on the results, then the analysis presented may not be appropriate nor truthful. A perhaps greater problem is known as data dredging or P-hacking, which is performing statistical analyses until the results indicate, statistically, what the researcher wants to "prove" in the study. See the case study at the end of this chapter on this topic.

How to securely manage large amounts of digital data for accessibility

Just about everyone agrees on several important issues with regards to managing copious amounts of data. First of all, data deposition should be done. The best time to do that in most cases is when the paper is published. This timing would also avoid the second issue: the current rate of data deposition is too slow. Third, data sharing must take into account security of human subjects, which is a separate, yet entangled ethical issue. Fourth, new models may be needed for data sharing. I described my DNA sequence deposition problem along with the problem of exponential amount of sequence data being generated and deposited into GenBank. The problem is extended to other -omics data including those from proteomics, metabolomics, ionomics, and glycomics, to name a few; then there is interactomics – how all the -omics interact with one another. Large amounts of data are coming from areas outside of biology, including Large Hadron Collider in Switzerland and large telescopes from around the world (Lynch 2008). More data will be generated in the next decade than from last century, maybe more than in the history of the world. Clearly, new resources are needed to manage it all.

It seems that -omics and biology are leading the way in developing new paradigms for data sharing. It is not an easy path. Let's assume that we can make and manage the storage capacity in the world for all the data we can possibly generate. Let's also assume that we believe it is a good idea to provide useful archives – researchers, universities, funding agencies, and the public all agree. It is clear then that getting researchers within a discipline or subdiscipline to agree on standards and protocols are doable. We have seen that in several occasions – this is what scientists do best (May 2009; Lynch 2008). The key here is the development of professional standards for data deposition to assure it is of sufficient quality and an appropriate format (National Academy of Sciences 2009). This probably has to be field-specific. Bioinformaticians and computer scientists can write routines to help search for and manage appropriate data. We've seen the scaling of certain old routines such as BLAST to include multiple processors or even supercomputers to allow faster and more complete searches (May 2009). It is also clear that universities and academic researchers are not the ones to make and maintain large databases; that should be left up to governmental agencies. There are instances where universities did set aside resources to maintain digital data, for example, the University of Rochester in 2003 (Nelson 2009). The cost of the repository was $200,000 USD, but if researchers refuse to use it then the resources are squandered. It is clear that large databases that are subscribed to must be supported by long-term funding as well. One can make the argument that maintaining data accessibility is at least as important as collecting the data in the first place.

Judge yourself

✓ Do you procrastinate or ignore tasks that are not pressing, such as data archival?
✓ Do you think the science community should be more proactive in demanding data archival? How?
✓ What may prevent you from simply cataloging all of the pertinent data at the time you publish paper; e.g., in supplemental files to the paper?

Data sharing and credit

One proposed way to handle massive amounts of diverse biological data is to create a wiki (in Hawaiian, meaning "quick") in the spirit of Wikipedia, where researchers work together in an open environment to manage and interpret data (Waldrop 2008). Wikis are, by definition, an altruistic mechanism to foster collaboration in an open environment. There are, for example, wiki projects to annotate genes – that is to put gene function together with its sequence. So, for example, as researchers uncover gene function information for the bacterium *Escherichia coli,* then they could add it to the EcoliWiki (http//:ecoliwiki.net). At the surface, wikis seem like an effective data management vehicle when researchers are all part of a very large project and need a common meeting place for their data or for researchers with a common goal or organisms. However, as Waldrop (2008) points out, this approach might not be sustainable because of funding and focus constraints. Providing organization for, or contributing to, a wiki does not directly lead to either funding or publications. These are the two endpoints scientists are most motivated to focus on. In addition, by their nature, it is also problematic to parse out credit on wikis. Therefore, wikis, while having roles in data management, might not be a cure-all for data curation.

There are various cultures among disciplines about sharing data. Some of the factors affecting sharing are historical and some are based on the nature and amount of data and collaborations. Certain fields such as astronomy and genomics have a strong culture of sharing data, whereas other fields such as social and public health sciences and climate sciences have a weak culture of sharing data (National Academy of Sciences 2009). Even if data sharing is mandated by funding agencies, there exists an honor system as such for adherence. Indeed, peer pressure to share or not to share is a strong force in specific professional communities of scientists. Some scientists wish to keep private data within large collaborations or milk as many publications out of a data set as possible; potentially setting up careers of graduate students or postdocs (National Academy of Sciences 2009). An important aspect, therefore, in the ethics of data sharing is one of attribution of credit. Think back to the concept of egoism as described in Chapter 1. Egoism is the ethical concept that people ought to do what is in their own best interests. If there is a short-term or long-term benefit to a scientist's career in sharing data and gaining publications and recognition by sharing, then chances are the scientist will share. The opposite is true as well (National Academy of Sciences 2009). One motive to incentivize the creation and feeding of datasets is to make them citable by issuing digital object identifiers (DOI), just as all publications have (Anonymous 2009; Thorisson 2009). Thus "data DOIs" elevate data sets to a standing otherwise not experienced currently – essentially making them like publications. Indeed, this is the way of the future, since scientists can now garner credit for generating and depositing large data sets that can then be utilized by a community of scientists. But as Thorisson (2009) points out, just because data sets are made citable does not mean they would be cited. In a wiki format, where various researchers add to a database from time to time, how would credit be divvied up? Unfortunately, there are more questions than answers right now with regards to data sharing, but I would like to end this section with a personal reflection, that also takes into consideration sharing not only data, but also materials.

Sharing science, a personal perspective

My research has long utilized various -omics data, but my lab mostly does synthetic biology and biotechnology. Whereas the genomics community is accustomed to deposition of data into public databases, the culture in biotechnology, which often generates more material than copious database-worthy data, has a different community perspective on sharing than does the genomics community. Biotechnology research generates gene constructs (e.g., plasmids) and transgenic organisms, to name two predominant materials, that other researchers periodically request. Once a paper is published and the materials are described, then it is fair game to give them away – usually under a materials transfer agreement (MTA) for others' to use in research. A typical academic MTA for people requesting materials to use in non-profit research is not very restrictive, containing with a few conditions. For example, the giver of the material might not want it passed from institution to institution without first notifying the source lab. The giver might want to know what the material will be used for. And the giver might wish for a certain paper to be cited when the receiving lab uses the material and includes its use in a publication. The point of an MTA is to pass along interesting and useful materials, but to assure a paper trail follows. When we enter the world of for-profit entities, involving either the potential giver or receiver, the MTA business gets a little more elaborate. As an academic scientist, I am generally happy to share materials and I appreciate when others' materials are shared with my lab. And as I get older, I've become even more eager to share materials. When I was a less experienced researcher, I harbored many questions about sharing materials and if I should share in the first place. Will the other researchers do something with the materials that I would otherwise get around to doing? E.g., will they scoop me? Will they use the materials for funding that I could otherwise win? Would I be enabling my own competition? Will they publish without my name on the paper? Will they make my research obsolete by doing all the good science first? Will they "misuse" the materials and publish science that will make me look bad? Notice that the questions were mainly about me.

At the same time I pondered these questions, I was continually asking several other researchers for materials. Sometimes they just sent it with no MTA and no questions asked! Wow! That made life really easy. But sometimes my requests were totally ignored. Follow-up emails then telephone calls went unanswered. That made my research life more difficult. In most cases, people shared their materials with me under an MTA, and that was fine. Doing the projects became easier with my group not having to reinvent the wheel every time, and I got to meet (at least over the internet) a lot of interesting scientists. At that time, I was also given myriad advice about sharing. In general my approach could be characterized as pretty cautious. I also worried about sharing my materials with two different researchers who were competitors between themselves – would that make one or both of them mad? And mad at me?

Looking back I can reduce all my worries, fears, concerns, and tribulations to a simple thought. I look back with a happy heart at many times that I shared materials and made sharing easy. I look back with regret those few times that I refused to share or gave someone the silent treatment, simply not answering their emails. Starting with the regrets: even though I had good reasons (or thought I had good reasons) not to share, I look back now with regrets that I didn't help someone I had the capability of helping. I violated the Golden Rule. I wish I could undo those instances. The outcome of most, but not all, the occasions I shared materials was positive. The materials were mostly helpful to people. Sometimes they even included my name on publications as they asked for my help with the contribution beyond the materials. I experienced new chances to collaborate with researchers I admired. Sometimes I was not included on publications I wish that I could have been. A couple times, the researchers gave my materials away to other people without my knowledge. Once or twice I probably enabled competitors to better compete against me. But in the end, I have no regrets about sharing – I would share all over again – I only have regrets for not sharing. I can reflect back on my career so far, and I conclude, from an egoistic point of view, that it is good to share materials and data. It is beneficial to share tricks of the trade and know-how. I arrive at the same conclusion from an altruistic point of view as well. As I think about how sharing might have affected my trainees who played key roles in developing materials I chose to distribute, I can't think of a single instance where there were negative ramifications on their careers or science. Today, I put information on how to acquire my more popular materials on my website – still in the form of an MTA with the minimum number of restraints mandated by my university administrators – but I no longer consider saying "no" for any reason a good reason with the exception that a prior MTA disallows it.

Judge yourself

✓ How easy is sharing for you? Do you enjoy collaborations and helping people?

✓ What are the barriers, if any, to your sharing materials, data, and expertise?

The land of in-between: ethics of data presented at professional meetings

The two extremes along the continuum of starting a project and publication are clear with regards to the propriety of data. At the start of a project, there's not a lot to talk about except concepts, ideas, and experimental design. The end of a project or subproject culminates in a journal publication and data deposition. The middle part can get a little tricky. When you have a partial dataset, how much of it do you show and discuss at scientific meetings? What about propriety of others' data shown at meetings? The combination of enhanced competition for funds and pages in top journals couple with advanced communication technologies lead to squishy ethical ground.

Why go to scientific conferences and workshops in the first place?

Scientific conferences can be standing annual meetings held by scientific societies, or ad hoc symposia, or special workshops to address a specific problem. The reason to attend these is to see and be seen, hear and be heard. From my perspective, conferences are a wonderful places to air ideas and data and enter into conversations with the smartest people in your field. Typically, many scientists in attendance either give talks or present posters, and then attend talks and poster sessions. The information presented can be a combination of published and unpublished data, but typically the latter is most interesting. Indeed, some meetings like Gordon Research Conferences specifically ask presenters to not present published data, i.e., old stories. Oftentimes the new data might be tentative with respects to conclusions. Scientists can "try out" their data to see if it passes an informal peer review of sorts that happens at conferences through normal interactions. Therefore, conferences are a combination of give and take – the Golden Rule once again. Often, talks or posters are qualified to be given at a conference by an abstract, which is a synopsis of the presentation. It might contain some data, but typically not very much can be said in a few hundred words. It serves as the advertisement and teaser for the presentation the public face. Meetings of 50, 100, 500, or 5000 scientists are usually rewarding in different ways. Disciplined scientists attend several conferences a year in order to stay current with the cutting edge of science. After all, few scientists have the opportunity to read all the important current papers in their field of interest. Why read them, when the papers to be published next year will be squeezed into a 15-minute talk with the highlights, or a square meter poster display. In addition, you get the chance to meet and discuss science with the authors. It is the place to meet future students, mentors, collaborators and friends. Professional meetings are time well spent. Just as one example of what can happen, I recall my postdoc's poster presentation where we had elaboratively laid out our grand plans with some supporting data to show that the big plan was feasible. Another scientist whom I'd never met pulled me aside and told me that his group had already tried most of what we're proposing and it didn't work. Furthermore, this scientist said that our preliminary data were artifactual and flawed. After choking on my coffee and doughnut, we had a great conversation and we had a productive collaboration for years. Not only that, my postdoc could focus on better experiments instead of wasting time on things that would not work. So, if meetings are such a great place to exchange ideas, what're the problems? To name a few, cameras, blogging, and tweeting, which are then countered by secrecy and withholding data because of a lack of trust. Technology interfacing with ethics are behind many of the current issues that hold the power to change the face of scientific meetings.

Meeting etiquette

When I present data at a meeting, I typically want the audience to see my best and most current stuff. I want to tell an interesting story during my oral presentation, so I'll cite my published papers by including the most pertinent data to tell the first part of the story in the setup. I'll also include information generated by other scientists.

For the climax, there will be my group's new data that nobody outside the lab has seen before. Taa daa (sound the flurry of trumpets)! I freely give my best story in its context to all those in attendance. Of course, there are caveats. Patentable information has to remain necessarily vague or unpresented so as not to jeopardize patent rights. If my collaborators also have sensitive information, I won't be able to share all those details there either in ethical deference to them. But generally, I don't try to hold back and I also try not to oversell (my group will chuckle at this one). There are several reasons for this approach. First, I want my peers to stay current with what my group is doing. They'll be among those reviewing my grant proposals and papers, so it is worthwhile to demonstrate competence and activity. I certainly don't want them to think that we're simply saying the same old thing meeting after meeting. Second, I treat the unveiling of the new data, especially those I'm not completely sure about (or their interpretation) as an audition. I want to gauge the audience reaction and listen carefully to underlying issues in their questions. Not too long ago at a meeting I excitedly presented results from an experiment performed by a graduate student. His interpretation of the data was questionable in my mind. I thought, sure, the reality could be exactly as the student suggested, but I questioned him about the possibility of other points of view being as likely to be valid. In the end, he assured me he was right. So, at the meeting, I played it exactly as I was coached and at the end of my talk, the hands flew up. I received very polite queries about a few things I stated, but the focus was on the potential controversy of our opinion of the data. After the session was over and as I mingled with a group of esteemed friends, there was one person who told me point-blank that he and colleagues thought I was wrong. I listened and thanked him – appreciating his forthrightness – then I proceeded to telephone my graduate student to relay the bad news given by the experts whose seasoned opinions I respected. Then, when we submitted the paper a few months later, we then had written the correct interpretation thanks to an honest and helpful scientist at the meeting who were willing to give me constructive feedback. This paper sailed through review and was published. Success for both the student and science! This is exactly what I strive to learn and achieve via meetings.

Now let's imagine this scenario going a different way. Let's imagine that I'm afraid to float my trial balloon of data interpretation – either because I'm afraid of appearing inept or because someone will steal my ideas or data. Then, the chances are higher that we get hammered by the peer reviewers of the publication a few months later. But that's ok too as long as we eventually get it right. Let's imagine that I present our data and my opinion at the meeting, but it is then "stolen" via camera – my slide is photographed – and then distributed via the internet with the caption, "Stewart's wrong in stating X." Then after I find out, I'm furious and then launch into damage control mode. These are two suboptimal processes that are becoming too common of occurrences in meetings. Of course, it didn't happen that way for me, but it is not beyond possibility nowadays. And this is making some scientists more guarded about what information they choose to present at meetings.

Some ethical questions are worth asking. For example, what is the accepted etiquette for oral and poster presentations if you are an audience member? Whenever I'm in oral or poster sessions, I regularly see camera flashes. I never thought much about what happened to all those pictures. I reasoned that people photographed slides or posters for their personal use. After all, I reason, that's not much different from taking notes, and everyone would agree that nothing's wrong about taking notes during a talk. Brumfiel (2008) tells a different tale. A group of Italian researchers had collected and tightly held onto satellite data that were gauged to be important by researchers in the field of astrophysics. They had presented slides with the data at multiple conferences. They had not yet published the data, however. Apparently, the slides became predictable enough for other researchers to quickly photograph key slides at a specific point in the talk. From those slides, the picture-taking scientists in the audience recreated data and published them first, citing the Italian group's presentation in their publication. On one hand, this seems very underhanded and sneaky. On the other hand, being slow to publish while simultaneously serially broadcasting data at meeting after meeting seems to be asking to get scooped. The Italian researchers were not very happy about the situation of indeed getting scooped. Another development in meetings is blogging, tweeting, and posting other's data presented at meetings on websites (Brumfiel 2009). Defenders of such actions justify blogging by saying that the whole purpose of scientific conferences is "sharing with the world what you're doing" (Brumfiel 2009). To them, it is not important if the person receiving the information is in the same room or not. I tend to disagree here. The spirit of scientific meetings is indeed sharing, but it is geographically and spatially confined sharing. The eventual published paper is intended to transcend space and time for wide dispersal; that is not really the function of a talk or poster. Conference presentations are often half-baked, after all. If blogging and tweeting my data becomes the norm at meetings, then I'll tend to present less "risky" data, and what's the fun of that? The Cold Spring Harbor conference organizers seemingly agree with my view in that they instigated a policy that requires bloggers or broadcasters to get permission of presenters prior to putting presented data on the internet (Brumfiel 2009).

I'm writing now from an airplane returning, coincidently, from a meeting in Italy. Last night, over a delicious seafood dinner and a few bottles of wine shared with old and new scientific friends, this very topic of meeting etiquette was discussed. How and when should information that is disclosed be used by attendees? One person says that once information is presented, it is public and anyone can use it – no questions asked. Others thought that an attempt to reference particular novel information should be made; that attribution is important. We all agreed that sharing ideas and data are enjoyable parts of meetings. There were no bloggers or tweeters at our table. But one of the scientists there told the story of someone basically publishing a book based upon lectures that had gone completely unattributed. The author led others to believe that the ideas were entirely the author's, and we all agreed that was an unethical action. We also agreed there are many gray areas, but there are some lines that shouldn't be crossed. That said, there are a few clear-cut decisions I make to accommodate certain situations. First, if my talk is conceptual and I plan on using the concept as the basis of upcoming grant proposals, I may not tell everything. After all, I do want to have a competitive edge in grant competitions. Second, if a talk is recorded like it was most of the time during COVID-era meetings when virtual conferences were the rage, I present fewer new data than an in-person meeting. After all, it is recorded. So, I take that into consideration. Honestly, I was not thrilled to attend or present virtual conferences. The third situation focuses on posters. In a single photograph, a person can capture my entire poster in seconds for future use. Therefore, I present nothing on a poster that I don't mind the world seeing.

Judge yourself

✓ How do you feel about dispersing information presented at conferences to others not in attendance?

✓ How confident are you in your own ability to make good decisions as a presenter and attendee of presentations?

✓ How comfortable are you in sharing unpublished data and ideas?

Raw data, processed data, and data analysis: ways to go right and wrong

Future of data

We live in an era of rapidly changing norms and practices about handling data – both before it is published and then afterward. Technology is at the root of these dynamics. First, we are able to collect more data in a single -omics experiment now than many scientists collected in their lifetimes. The challenges to handle and archive it all are immense. Second, the internet has made accessing digital data easier than ever enabling scientists to quickly form virtual collaborations or perform meta-analyses. Therefore, it is desirable to place datasets in accessible locations and in useful formats. How to actually accomplish these efficiently are not clear. Third, meeting etiquette seems to be changing as photos and blogging of preliminary results can be dispersed quickly and widely on the internet. All of these issues are certainly subjects of debate for years to come.

Summary

✓ There is a level of trust required when sharing published or unpublished data.

✓ How and where data are shared and archived are very important issues that are worthy of debate as science and technology continue to evolve.

✓ Good stewardship of data is one of the most important parts of science.

Case study: producing the answer you want

Percy D. House was a postdoc in a plant breeding lab run by a supervisor that advises postdocs that they should produce at least one paper per year to be competitive for an eventual permanent faculty or industry position. In a relatively small growth chamber study of clonal plants bred for a range of fruit size, Percy collected a lot of spectral data of plants grown under three different LED light treatments. He reported using a Fisher's LSD multiple comparison test after an analysis of variance to analyze the spectral data. Percy hypothesized that a number of hyperspectral bands of plants grown under different LEDs would lead to an early prediction of fruit size of plants grown in the field. Of the many bands analyzed, Percy reported that one reflectance band grown under blue LEDs was predictive of statistically significant fruit size in the field. The study was deemed to be biologically important, even though it was small, since the approach could be used to speed up plant breeding cycles. After the study was published, Percy abruptly left the lab because he was hired by a company whose scientists were impressed by the paper, which was authored by Percy, a technician (who was also the lab manager) who helped make the crosses and grow the plants in the field, as well as the supervisor who got the funding and designed the study. Percy did all the growth chamber experiments, data collection, and data analysis. Two weeks after Percy had left the university, the lab manager examined Percy's university-owned lab computer and found the original spectral data along with the statistical analyses. She was surprised to find that Percy only analyzed data from one hyperspectral band (out of the hundreds captured), and then had used a wide variety of multiple comparison tests, even though the ANOVA did not show significance in the model. Only the Fisher's LSD test showed statistical significance, which was then reported in the paper. In the paper, Percy had claimed that all the bands were analyzed for significance at $P = 0.05$, and he made a convincing argument why that one particular band should predict yield. The lab manager tried to contact Percy to answer questions about the data and analyses, but Percy never responded.

Discussion questions

1. Does it appear that Percy was responsible for hypothesizing after results are known (HARKing) and/or P-hacking?

2. Should the lab manager report her findings to the supervisor and/or the university official who oversees research misconduct cases?

3. Should the lab manager and/or the supervisor alert the journal editor and request the paper be retracted?

4. Is there anything the supervisor and/or lab manager could have done to prevent the problems?

5. Should Percy's new employer be alerted of the suspicious data analysis and situation about Percy?

Case study: who owns the data and notebooks?

Data are usually ultimately owned by the performing institution or company. This was a difficult lesson learned by Judy Mikovits in 2011. In 2007 the Whittemore Peterson Institute for Neuroimmune Disease (WPI) in Reno, Nevada, hired Dr. Mikovits where she became research director to study chronic fatigue syndrome (CFS), one of the focal points of WPI's mission. Dr. Mikovits became famous for a study published in *Science* linking a mouse retrovirus to CFS in 2009, and then infamous as the findings of the study crumbled under peer scrutiny and retraction of the paper in 2011. Independent of the retraction were issues at WPI that led to her dismissal the same year. In an unusual move, Dr. Mikovits "stole" a WPI laptop computer, USB drives, and lab notebooks and took them back to a boat near her home in Ventura, California. She hid on the boat to avoid a summons (Cohen 2011), but was eventually arrested for breaking in. The charges were eventually dropped, but she spent five days in jail. At the crux of the matter was intellectual property on the computer and items that were removed from WPI. In the end WPI got back the stolen property, and Dr. Mikovits' career took various odd turns that most scientists would loathe to imagine. She hasn't published any scholarly papers since 2012. She is now more closely associated with conspiracy theorists than science. During the

COVID-19 pandemic she made several claims including masks can harbor coronavirus and make people sick. These claims have no scientific support. She has been a critic of Anthony Fauci, vaccines, and star of the short film "Plandemic: the Hidden Agenda of Covid-19."

Discussion questions

1. What is intellectual property?

2. Why would institutions typically own intellectual property and the media containing data that is entwined in intellectual property?

3. What went wrong with Dr. Mikovits' career in science? Or was her path from a PhD in 1991 to COVID-19 critic acceptable?

Chapter 12

Conflicts of Interest

ABOUT THIS CHAPTER

- An active and productive research program may create a number of potential and real conflicts of interest and commitment.
- Conflicts of interest and commitment can be particularly complicated for university scientists.
- In addition to being highly productive, successful scientists find ways to manage or avoid conflicts of interest and commitment.
- It is important for scientists to be transparent about conflicts to their employers, funders, and the community-at-large.
- Undeclared dual employment and funding situations can jeopardize careers.

A superb example of ethics transcending morality can be seen as we look into conflicts of interest, which can occur when someone has more than one interest in which one has the potential to corrupt the other, for which the latter is typically a constraining interest. The underlying values, as we consider conflicts of interest, are loyalty, priority, and then transparency. If a person is loyal to a certain entity or cause, oftentimes the person must say "no" to other opportunities or causes that result in conflicts with the first. For example, marriage is widely considered an exclusive romantic relationship. Married people are wise to forego other romantic relationships given the conflict of interest. We see other examples of needing to avoid conflicts

---

Research Ethics for Scientists: A Companion for Students, Second Edition. C. Neal Stewart, Jr.
© 2023 John Wiley & Sons Ltd. Published 2023 by John Wiley & Sons Ltd.
Companion website: www.wiley.com/go/stewart/researchethics2

of interest in all walks of life. In the judicial system, a lawyer cannot represent both the plaintiff and defendant, since the two sides are in conflict by definition. A gang member cannot be both a Blood and a Crip, since they compete with one another. A professor has more latitude than lawyers or gang members, but conflicts of interest can occur. Sometimes they should be avoided and other ones can be managed. Lines of loyalty are valued, and it matters not whether the cause or morality in the situation is good or evil. Loyalty matters. In science, the overriding primary loyalty is indelibly connected to discovering truth in the natural world. As we saw in Chapter 4, Herman's Second Law is "In research, what matters is what is right, not who is right" (Herman 2007). Scientists strive to uncover and truthfully explain phenomena, causes and effects without bias. It is the potential introductions of biases, both subtle and blatant, that are most worrisome to scientists. Thus, dogmatic ideology and science are not good bedfellows. However, most often conflicts occur when money is involved and not ideology. For example, if a university researcher performs research on a company's chemical while being paid by the company to do so, then there is an inherent conflict of interest. What could such a scenario mean for the subsequent publication of results? Employers of scientists must also be vigilant to guard and manage for conflicts – both for sciences' sake, but also for fiduciary reasons to protect both institution and their employed scientists. Of course, it is also expected that we should have solid loyalty to our employers. We must be careful not to inflict injuries to this relationship by taking on pay from others that could be construed as a conflict. Finally, the ultimate conflict comes when we compromise our consciences.

The dynamic landscape of conflicts of interest

Today's science differs from that of yesteryear's by the sheer quantity and breadth of possible conflicts of interest and conflicts of commitment. These are often created by commercial interests in academic scientists and their science, especially in lucrative translational and applied science (Louis et al. 1989). Although conflicts of interest abound and are probably unavoidable in certain fields, most of these are manageable. In most cases, outside consulting, business ventures in startup companies, patents, and the like can actually be beneficial to a scientists' home institution and its students. But on the other hand, things can go horribly wrong and out of balance when a scientist crosses a line of demarcation in the world of conflicts of interest. Case studies will be examined that involve several layers in which a scientist is forced to make decisions that have serious downstream ramifications. Potential student reactions and the effects of an out-of-bounds mentor will be discussed. We will also discuss faculty members being employed by more than one university and in different countries: having shadow labs and how these must be disclosed to funders and to the primary employer. We have already covered conflicts of interest regarding co-authors and collaborators and how they pertain to reviewing their grant proposals and manuscripts, so this will not be discussed in this chapter. Rather, potential conflicts of a less straightforward nature will be discussed.

I will focus here on conflicts of interest and conflicts of commitment with regards to university scientists, since institutions generally have fewer rules (and greyer rules) than industry and government in this regard (and there are good reasons for it) and there are many levels and instances where conflicts can crop up. Why should we not worry about conflicts for industry and government employees here? For industry scientists, the rules are clearly understood that the conflicts world is black and white. The company that employs a scientist expects no outside professional interests to exist, at least none where there is opportunity for payment or loyalties to be potentially compromised. Whatever is done at the company by the company is owned by the company. In many ways, the situation is the same for government employees. A few years ago, National Institutes of Health employees could perform outside work as consultants, but that is no longer the case, because of conflicts of interest that ensued could seriously affect the NIH to accomplish its mission. The relationships that industry and governments have with their employees have high fidelity – no chance to play the field. In exchange, compensation and/ or employee benefits are motivators to remain loyal and conflict-free.

Potential conflicts of interest for university scientists

Let's clear up a potential misunderstanding for the lead-in discussion about the differences between industry/government and universities. I'm not advocating that it serves universities or science for faculty members to be disloyal in any way. And I don't think that professors ever want to be seen as disloyal to their employers. I think that sometimes it is simply a slippery slope driven by prospects of earning additional money, finding more research funding, or the pursuit of compelling outside interests that cause faculty to find themselves in conflicts of interest dilemmas. In some cases, maybe most, these pursuits can be beneficial to professor, student, university, and society. Today's universities want their faculty to be interesting, in demand, and entrepreneurial, but they also must maintain appropriate boundaries to prevent the university and objective science missions from being compromised. To execute a university's mission of educating students, providing service for stakeholders, and creating knowledge, it is sometimes advantageous to enter into agreements and situations that are somewhat muddy. The over-riding values that must be elevated to the top are honesty, transparency, and fair play. All relationships should be disclosed or disclosable at the appropriate time. I use the "publicity rule" as my own guide. That is, if a relationship or deal involving me were to show up in the news or on prominent social media, how would I feel and how would it affect my professional reputation? Would I feel embarrassed and wish to hide it from my employer and colleagues, or would it be the reverse? Conscience is a good guide, and so is the publicity rule.

I am currently a professor at the University of Tennessee. I teach a few courses and run a research program that has been funded by US federal agencies, companies, foundations, an endowment, and various other university funds. By far, most of my research funds come from US government agencies. Funders and the university officials expect me, first and foremost, to be productive in these educational and research pursuits. I share these expectations and values with them. University administrators would not be thrilled to if I had a dramatic drop-off of funding or publications. They want assurances that students are benefiting from my teaching and research programs. Staying productive and close to the cutting edge of science is also important for my career and competitiveness. Thus, we have a good match with shared expectations. That said, I've also founded and participated in university start-up companies using university-owned intellectual property. I've been paid as a consultant and expert witness by companies. I've been paid to speak and serve on grant panels and government agency scientific advisory panels. I have served as a member of a scientific advisory panel for a company. Are any of these examples conflicts of interest or commitments? Are they allowable or should aspects of these activities be mitigated? The big questions: Are they good for universities and for science and society? Is there potential for harm? Is any harm realized? These are all important questions for me to ask myself and the university to ask as well. We'll examine what constitutes and defines conflicts of interests and commitments and try to grade them on severity. We'll also grade their risks and benefits. One handy tool that I'm using as an outline and starting point is my university's outside interests disclosure form. Most colleges and universities use these instruments to flush out yearly disclosures about faculty activities and compensation outside of the institution for each of their faculty and staff members. Collectively among institutions, university lawyers, and administrators have thought a lot about what potential conflicts should be disclosed. One driver for requesting disclosures is that in every university there have been cases of faculty members who have gone overboard on certain outside activities or who have certain interests have embarrassed or compromised the university and its mission in the past.

Conflict of interest vs. conflict of commitment

Conflicts of interest deal with the potential of a secondary cause to corrupt a higher, primary, or constrained cause. We can think an outside activity as a potential source of pollution. They are usually defined as professional vs. private interests where someone could have some personal or financial gain outside of their primary professional duties that could impair university performance or corrupt scientific objectivity. A conflict of commitment might have more to do with an outside time or personal energy allocation that could serve to diminish full focus on the primary mission of a scientist. One can imagine several instances where there would exist conflicts of commitments, but with no conflicts of interest. For example, deeply investigating and spending one's time doing pro bono scientific research for a good cause might incur a cost to one's primary job – that is teaching and doing funded research to support students and university science. In this case, there would be no personal or financial gain, but it could be an inappropriate time sink that would distract a professor from his main duties of research and teaching. These two entities, conflicts of interests and commitment tend to get rolled into one entity for practical purposes. Below we'll cover the main conflicts that university scientists face.

Outside positions

Most universities are not keen on their faculty having political or for-profit positions outside of their institution. For the latter, some are typically tolerated, but there must be clear delineation between duties as a faculty member and duties in the private sector and how time commitments are to be managed so that faculty duties are not compromised. Typically, institutions look at three things. First, they want to know how much money is involved, say, in terms of salary per year if it is greater than some defined threshold, say, $10,000 per year. Second, they want to understand how much time commitment is required for the outside interest in terms of days per month. For example, there is a difference in becoming the CEO of a company that has one employee (you) that deals with trading baseball cards, a pursuit that can be accomplished on the weekend at home, and being the chief technology officer of a start-up biotechnology company that has 10 employees. This leads us to the third concern: the particular area of business. Universities are not keen on being competed against by their own employees. Neither do they relish losing a valued employee to the private sector.

Consulting

Many university professors do consulting for companies in their area of expertise. This is typically a win-win-win for universities, faculty members, and companies. Institutions become concerned when the sheer volume of consulting becomes too great or the annual income from a single consultant goes over a certain threshold. Typically, universities will allow professors to consult, say, two to four days per month as part of their official duties, leaving the exact details up to their individual professors. They view this moderate amount of consulting as a worthwhile and manageable endeavor. For example, because of my own consulting work, new contracts and grants have been initiated for the university and students have been provided opportunities for employment and professional and scientific growth. Improving economic development and impact is part of a university's mission. Of course, if the compensation for consulting becomes too high, university administrators might question loyalties and the ability of a faculty member to sustain objectivity. They might also question whether there is a conflict of commitment.

Boards

Being on a scientific advisory board of a company or steering committee for a nongovernment agency, or even a panel for a government granting agency is seen as beneficial activities for universities and faculty alike. Many of the same reasons apply as seen above for consulting. However, serving on boards are viewed as more long-term commitments, and the rules of engagement are somewhat different because of potential longevity and depth of involvement. I personally have no worries about having consulting agreements with two or more companies in the same research discipline, even during the same overall time span (but not on the same days). Companies typically request I sign a confidentiality agreement that entails my not sharing compromising information with anyone else. I bill by the hour and so my loyalties to each company, time wise, are fixed, discreet, and discrete. Being on a board is both a deeper commitment, with regards to information shared and it is typically also of a longer duration – years instead of days for consulting. It is typically also made public from the company's perspective. Therefore, the professor must be careful not to develop conflicts of interest with regards to serving on two boards for companies who are in the same field. Boards can also be tricky with regards to the amount of time needed to serve; it can potentially conflict with time needed to do research and teach. But again, boards can lead to new synergistic opportunities that can benefit universities. Many times students, even undergraduates, eventually start their own technical or scientific companies, whereby they tend to contact a trusted professor to consult and later serve on a board of advisors. Universities welcome the maintenance of contacts with their alumni who have a sense of loyalty to the institution. Again, it could mean funding and job opportunities for current students and increased visibility for the institution.

Start-up companies

Start-up companies are usually founded by faculty members to develop products stemming from research performed in an academic lab and perhaps patented by the university. A professor, typically with tacit approval by the university, might judge that a quicker and better path to economic development and commercialization would be to create a company rather than simply out-license the invention to an existing company. In this way, the inventor can shape a technology's development in a stronger way. There are other reasons to found companies too, including potential sizable financial benefits to faculty members, universities, and their surrounding communities. In fact, entire industries can be founded and regions of excellence by university start-ups. The San Francisco Bay and Boston areas are two examples of biotechnology hot-spots as a result, at least in part, from companies co-founded by faculty members, e.g., Genentech and Biogen, respectfully. Universities can own equity in companies and stand to greatly benefit if companies succeed. But at the same time, start-ups are also potential pit-traps for conflicts. For example, a conflict of interest might arise if the professor chooses to have line management duties – for example as the CEO. It is cleaner if the professor has scientific or technical oversight duties in which there are no people-management responsibilities; that is, the scientist is advising the science. Professors, or even students, having equity stakes in start-up companies are ok. It is fine for professors and students to benefit from patents. Conflicts of interests come in to play when the equity stake for the faculty member (and university) is too great; even far less than controlling interest can be considered too great. For example, Stanford will never own any more than 10% of any of its start-up companies. Perhaps the bigger potential problem surrounds conflict of commitment. Start-up companies are a lot of work to launch and to sustain. If the faculty member is spending so much time executing CEO duties, then it is not hard to imagine teaching and research responsibilities taking the back seat. There is also the potential problem of student or postdoc exploitation. Students paid from university grants should not be expected to work for the start-up. There should be clear lines of delineation and it is smart to keep department heads and deans in the loop with regards to managing conflicts. Not only are overlapping fields of interest important considerations, there needs to be demarcation of workplaces. In no instance should a faculty member use his or her university lab to run a company. Lab space should be rented.

Employment at more than one university and running "shadow labs"

In the past decade or so, various countries have funded so-called talent programs to entice established scientists from other countries to work at their universities and run laboratories in the same discipline as in the scientist's primary institution. The goal typically for both the recruiting university and the person being recruited is dual employment. By far, China has run the largest talent program and has paid large amounts of funds (say, over $100,000 per year) for a few weeks of work in-country. The expectation from the second university is for the professor to list their second university as an affiliation on papers and raise funds for research at the second university. Given that most US professors are on nine-month appointments in their primary university, they could potentially take second jobs over the summer during their other three months, if their time is free from grant constraints. Thus, the case can be made that no conflict of interest would exist. But, several highly publicized cases, some involving criminal prosecution, have centered around undisclosed dual employment. In some cases, the US government accused faculty of espionage, illegally sharing intellectual property, and violating export control laws. Only a few of these cases in the United States have resulted in guilty verdicts. One prominent example was former chair of the Department of Chemistry and Chemical Biology at Harvard University: Charles Lieber. He was convicted of six felonies including making false statements and tax fraud. Essentially, he lied to Harvard about his affiliation with Wuhan University of Technology in China, failed to declare income, and pay taxes on income. More often, aggressive charges by federal prosecutors have not led to guilty verdicts in defrauding the government by not listing secondary employment (Mervis 2022). Lesser charges and lenient sentences have been given or charges have been dismissed. While many cases may not result in felony convictions, faculty members have been dismissed from their universities for not disclosing and not being transparent about having second jobs (conflicts of interest and commitment), and not disclosing funding from other countries on grant forms (Adams and Schwendinger 2020). Thus, in many cases greedy motives have not worked out for scientists as they sought to essentially potentially double their funding in shadow labs, increasing their pay checks, and being an author on significantly more papers. I'm a bit perplexed about why they thought it would not end badly, but in most instances did. I recall being recruited by a number of international institutions for faculty jobs. In some cases, the expectation was that I would leave my current position to join another university, i.e., move in-country full-time. But in others, clearly there was an expectation that I'd run a second lab overseas and get paid quite well doing so while I maintained my US faculty position. I could never imagine a way that such a scheme would be holistically beneficial, and that it would likely be either a headache or a bona fide disaster. So, while I briefly explored some of the full-time offers, I declined to be recruited by the ones expecting me to run a shadow lab abroad.

Financial interests

Universities might be concerned about outside financial interests that are related to a professor's expertise. There might be a temptation to use the university position for large financial gain or even for the exploitation of university students or employees. Let's say that I have sizable stock holdings in an agricultural company that does business in my particular area of expertise. It is not inconceivable that I could be tempted to alter data or shade findings in my research on a block-buster insecticide to drive up stock prices to my direct financial benefit. Most professors make it a point to stay away from these situations because of this exact sort of risk of conflicts. Faculty member reputation is the foundation of their credibility in their fields.

Family's interests

Most of the above items also apply to the interests of a faculty member's immediate family: spouse, children and parents. Again, this is a precaution to stave of temptation of direct gain because of corruption.

Judge yourself

✓ Are you particularly susceptible to temptation or greed? Are you especially motivated by money?
✓ Do you have diverse interests that could be targets for potential conflicts of interest or commitment?
✓ Some people have idealistic delineations that lead them to completely separate science and universities from companies and profit-motives? Do you hold these ideals?

Conflicts of interest within labs or universities

Naively and idealistically many of us enter into labs and universities holding pure scientific interests, ideas, and ideals. Given that there is a good bit of stochasticity governing where we study, with whom, and what science we decide to pursue, there are many choices about the details of what we do in science. That said, during the early parts of our training and careers there are not many opportunities for the conflicts of interest to occur that are described above. Graduate students, postdocs, and assistant professors are innately focused on graduating, finding a permanent job and getting tenure, respectively. There is not a lot of time for outside activities. But the extreme focus and the need for high research productivity can set people up for internal conflicts of interest. Thus, it is generally not a good idea to directly compete with your own lab members or people at your university. Or your boss! Or your ex-boss. Or poach their employees. Yet these kinds of competitions seem to happen as frequently as best friends discovering that they often dress like each other. It is almost as if randomness is not so random after all. In the pursuit of publications and success, these internal conflicts are not very rare, but it takes energy to avoid or manage them.

There is no right answer to the dilemma of finding that you are in direct competition in science with someone that would ideally be a collaborator. The two people you definitely don't want to compete with are your advisor and former advisor. This is like dating your best friend's old girlfriend. It almost always ends badly with ill-feelings that can compromise both party's effectiveness in science. Many mentors are happy if their trainees continue to work in the same field, but others are more protective of their specific areas.

Beyond the advisor, there are many gray areas. The optimal, although at times idealistic and not feasible, approach, is to face all potential competitors directly and lay your cards on the table. Hopefully, the competitors will do the same. I've formed collaborations with former competitors. I have also decided to not pursue research that I knew might put me in a sticky situation with someone I respect and admire. Again, there is no one right answer. The best thing to do is approach a situation with integrity and dignity and to do what's best for science, your interests and career. Think long-range in these matters, and not just about the degree or next paper.

Another way to handle impending competitions is to *not* speak about the topic of interest to people who you think might be working along the same lines of research. In this way, not knowing the all the facts prevents you from being "contaminated" with ideas that could be used unethically. There have been a few instances where I've decided to guard against contamination rather than putting all my cards on the table. It was not that I was afraid of being scooped. It was that I was concerned about knowing what someone else was doing that could intractably affect the course of my own project, and I didn't want to chance it. Here is an example. I participate in a large project in which a key early activity in the grant was to decide which genes to investigate with regards to function for cell wall biosynthesis with an eye toward wall degradation. I was invited to review several grant proposals on this very topic during a key period of time when our project was in the early stages of deciding which genes to study. I declined to review all proposals having to do with cell wall genes. Why? They would have invariably shaded my judgment one way or the other. I could have been tempted to use the information to benefit my big project. But my primary concern was for the PI of each grant proposal I'd review. If I were the PI, I wouldn't want to provide crucial early information to a huge center that could potentially scoop me. My other concern was for the center. It seemed to me that the fairest way to play this situation was to remain uncontaminated. Once the center had its big gene list produced and research underway I felt I could return to my normal activities of reviewing related proposals without fear of inadvertently acting inappropriately in what could have been a no-win situation. One of the biggest ways to not find yourself in a conflict is to anticipate a potential conflict and avoid it.

Judge yourself

✓ When it comes to ideas and projects do you have almost compulsive or obsessive tendencies that could shade your view toward unhealthy competition?

✓ What is your style with regards to confrontation and compromise? These are important conflict management tools.

One last issue to discuss is intralab competition. When the PI is narrowly focused on a particular topic, there will invariably be aspects of the research that overlaps among the interests and pursuits of graduate students and postdocs. The goal for lab personnel and especially the PI is to minimize these overlaps. However, some wayward PIs (yes, alas, I've done this before) assign the same project or close-to-the-same project to more than one person as a bet-hedging strategy or to set up a "friendly" competition with the goal of completing a project at seemingly a faster pace than if one person had sole ownership of the project. The winner is then feted with first authorship. This play often turns out bad because there is a loser, and the loser isn't always happy. The loser's friends might not be happy either. In fact, the entire lab (even the winner) could be unhappy because of the lousy morale that ensues during and after the competition is completed and the winner collects the reward. Therefore, intralab competition should be minimized for the sake of total productivity and integrity of management. The PI has the most control of this situation to assure that most, if not all, lab members are winners.

Case study: the case of the promising new drug

Courtesy of Ruth L. Fischbach, PhD, MPE and Trustees of Columbia University in the City of New York; www.ccnmtl. columbia.edu/projects/rcr (Fischbach and Plaza 2003)

Dr. Linda Roberts has spent the past five years working on a new drug for the treatment of lupus erythematosis. The molecule she designed links a fragment of an anti-inflammatory drug with a protein that binds to the diseased cells. Designing this new drug was made possible by two decades of research in the Immunology Laboratory at Westfield University Medical School, where Dr. Roberts works. Without the basic work in researching the molecular biology of this disease (the early stages of the research were funded by the National Institutes of Health), the highly specific drug would never have been developed.

At the same time that Dr. Roberts' research has yielded such promising results, federal financial support for biomedical research has declined. If this new agent were an effective treatment and a commercial success, it could be extremely helpful for Dr. Roberts, her department, and especially the medical school.

Dr. Roberts' research in the past five years has been supported by funds from Arthrid, Inc., a company that markets a number of drugs for arthritis. She was also given a consultant fee of $50,000. Indeed, researchers working for Arthrid helped with methods for producing large amounts of the therapeutic molecule. Without the resources of a pharmaceutical company, developing a marketable product would have been extremely difficult, if not impossible. Also, changes in federal regulations governing research encourage collaborations between academic scientists and companies in order to promote the transfer of technology from the laboratory bench to the clinic. There has also been a trend for institutions to hold equity interest in the start-up companies of their faculty. Because Arthrid is a local company and has been generous to the medical school, several members of the Westfield University Hospital Institutional Review Board (IRB) have bought stock in the company.

This long-standing relationship made it feasible for Arthrid and the medical school to enter into an agreement that entitles Arthrid to own the patent rights to all discoveries made in the course of the research it funds and entitles Westfield University to 5% of Arthrid stock and a 5% royalty on sales of all products that result from the research.

In the highly competitive pharmaceutical industry, companies like Arthrid, Inc., seek patents on all promising discoveries. A patent gives the patent holder the right to exclude anyone else from making the patented product during the 20-year life of the patent. During those 20 years, the patent holder would hope to earn enough revenue from the product to recover the typically enormous costs of the basic and clinical research that leads to the production of the product. It is possible for the research and development of a drug, and the subsequent approval by the FDA, to take as long as 10 years and cost more than $800 million.

Included in the Arthrid-Westfield agreement is a nondisclosure agreement, whereby Arthrid will be able to protect its proprietary interests. There are restrictions regarding publication, including Arthrid's right to review all data and a mandate for Dr. Roberts to send to Arthrid all manuscripts at least 30 days prior to their submission for publication. This would allow Arthrid to delete any information that, according to the company's directions, should not be published or presented, which might threaten its rights to any patentable invention.

Extensive use of the experimental drug in animal models of lupus has been highly successful, producing the desired anti-inflammatory effects. Other similar drugs have been used without any serious toxicity. Thus, the drug is now ready for Phase I clinical trials. To encourage this collaboration, Arthrid would be pleased to pay for a trial at Westfield University Hospital. By all accounts, Dr. Roberts would be the ideal clinician to conduct the trial, because of her intimate knowledge of the drug. Arthrid is willing to issue to Dr. Roberts 2% of its common stock. In addition her husband will receive 2% of Arthrid stock and her 14-year-old son will receive 1%. If the trial is successful, this would go a long way toward covering her son's college tuition.

Dr. Roberts and her colleagues work in the hospital's Medical Clinic, which has a large number of patients with lupus who would be available and thus easy to recruit for the clinical trial. Also, because of the department's reputation in research and patient care, Dr. Roberts would be able to enlist the cooperation of other hospital departments around the country in initiating a multicenter clinical trial of the new agent for treating lupus. Dr. Roberts and her colleagues submitted a proposal to the IRB which they believe justifies the use of their clinic patients because of the benefit that this new drug will provide.

On the assumption that she will be conducting the trial, Dr. Roberts approaches her postdoc, Dr. Henry Chung, to ask if he would join her in testing the new drug. But Dr. Chung has been pursuing a different and potentially significant project, cloning a gene for asthma, and is close to completing his work on it. Completing the asthma project would put Dr. Chung in an excellent position to apply for a faculty post and qualify for a grant under a newly announced federal program; working with Dr. Roberts means that he must set this work aside. Dr. Roberts tells Dr. Chung that if he joins in the Arthrid project, the company will issue him shares of Arthrid common stock equal to 2% and also pay him a generous consulting fee. In addition, Dr. Chung would continue to receive his postdoctoral stipend.

Dr. Chung is already somewhat annoyed that Dr. Roberts is spending so much time at the Arthrid labs; he is not receiving the supervision for his asthma project from Dr. Roberts that he feels he needs. Westfield University Medical School allows faculty members to spend 20% of their time on outside commitments, and Dr. Roberts is spending about 12–15 hours a week at the Arthrid labs. Since Dr. Roberts works closer to 60 hours a week (she always works on the weekends and takes work home every night), she does not feel that her time away from the medical school is excessive. Furthermore, this time away from her medical-school lab allows her to work on the therapeutic molecule using laboratory equipment at Arthrid which her own lab lacks.

Meanwhile, Dr. Frank Bonita, a colleague of Dr. Roberts', asks Dr. Roberts for a small quantity of a reagent that has been used in the lupus drug research. Drs. Bonita and Roberts were students together, entered the department at the same time, have openly discussed with each other all their research for many years, and are good friends. Each has been indispensable in the research success of the other. The secrecy covenant in the Westfield-Arthrid contract now prevents Dr. Roberts from granting what would otherwise be Dr. Bonita's routine request for a reagent. Dr. Bonita wonders whether Dr. Roberts acted prudently is so restricting herself.

The IRB will meet soon to review Dr. Roberts' proposal to study the new drug at Westfield University Hospital. The IRB chair has been informed by the dean of the medical school how important this proposal is for the medical school and makes the IRB members aware of this.

Discussion questions

1. What is a conflict of interest?

2. Why does a conflict of interest matter? Why should the university be concerned?

3. What types of conflicts of interest can you identify in this case?

4. Should Westfield University Hospital undertake the clinical drug trial? If so, should Dr. Roberts participate?

5. Does it matter if Dr. Roberts' financial interest in Arthrid consists of consulting fees, or common stock (equity), or both?

6. Should Dr. Roberts recommend to her patients that they enroll in the clinical trial if it is carried out at Westfield University Hospital? What about elsewhere?

7. Is Dr. Roberts being faithful to her obligation to provide an educational experience for Dr. Chung?

8. Is Dr. Roberts acting properly in the way she chooses to allocate her time? Is this in violation of Westfield University Medical School policy?

9. How should Dr. Roberts respond to Dr. Bonita's request for the reagent?

10. What are the implications of the Arthrid–Westfield nondisclosure agreement for academic freedom?

Case study in review integrity: undisclosed conflict of interest

By Mike Lauer, NIH, published in the Open Mike blog, 2019 and used with permission

Sometimes it takes detective work to unearth attempts to undermine the integrity of peer review.

Take the case of Dr. Smith, one of the reviewers on a study section in the Center for Scientific Review. The scientific review officer (SRO) would like Dr. Smith to review an application with Dr. Jones as principal investigator (PI).

In checking for potential conflicts of interest (COI), the SRO cast a wider net and found something troubling. Dr. Smith, one of the reviewers currently set to review the application listing Dr. Jones' as PI, had been listed as one of the key personnel on an application with Dr. Jones as PI that was under review in another, recent study section.

It was obvious Dr. Smith had a clear COI as a reviewer for the application with Dr. Jones as PI. The COI instructions for reviewers state that a reviewer may not review certain applications and must leave the room when the reviewer, within the past three years, has been a collaborator or has had any other professional relationship with any person on the application who has a major role.

In this case, Dr. Smith, who is being considered as a reviewer for the application, is a professional associate of Dr. Jones, the PI on the application. However, Dr. Smith had not declared a conflict with that application.

The SRO immediately notified the review chief, who unearthed more information when searching PubMed. They found that Drs. Smith and Jones co-authored multiple research publications within the past two years. Coauthoring publications within the past three years also is a clear conflict of interest (**NOT-OD-13-010**).

The review chief alerted the research integrity officer (RIO), who found more irregularities. Turned out that Dr. Smith was on the study sections that reviewed a few more grant applications listing Dr. Jones as PI. At no time had Dr. Smith declared a COI with the applications. The RIO alerted the NIH Office of Extramural Research (OER).

OER terminated Dr. Smith's service in peer review indefinitely and the application listing Dr. Jones as PI was reassigned to a Special Emphasis Panel for initial peer review.

All participants and stakeholders in the peer review system are responsible for its integrity. Whether you are a reviewer, PI, or NIH staff, each of those roles is crucial. NIH is paying attention, and NIH is taking action.

This scenario is fictitious but based on real events.

Case study: the case of the crowded room – who should investigate a line of research when many people have an interest in it

PhD student Ginger Gonzales was recruited to a large university in the Midwestern United States to work in the lab of Dr. Freddy Pinto. Dr. Pinto is a productive PI, having multiple grants from the NSF and NIH. In addition he is a Howard Hughes Medical Institute Investigator, which gives him sizable funding per year to pursue essentially any line of research he wishes. Dr. Pinto sees great potential in Ginger and he recruited her to his lab to develop and pursue the project of her choice. Dr. Pinto's interest is broad, but he is best known for applying ecological principles to epidemiology and using various organisms to model and better understand epidemiological patterns. Ginger decides to focus on plants and ephemeral epidemiological phenomena in a relatively simple high latitude peatland system. She also wishes to incorporate global climate change parameters into her modeling and empirical experiments. Since this is somewhat out of Dr. Pinto's direct experience (although he is fascinated with the potential of the project), he encourages her to speak to several faculty members within his department of biological sciences and also in other, more applied, departments.

As Ginger talks with many people at the university about the project, she gets helpful feedback and there is a great willingness from faculty to help her. They also are encouraged by her initiative and ideas. One way faculty can help students is simply in engaging in scholarly and open discussion about ideas. Another way is volunteering the use of lab equipment and materials with few-to-no strings attached. A third way is to volunteer to serve on her committee, which a couple of professors do. She contacts scientists in other universities and also gets cooperation and good ideas from some, while a few others are more standoffish. When she asks Dr. Pinto about the different responses, he tells her that most scientists, especially close colleagues, respect Pinto and his group and are eager to see science progress. They especially enjoy helping students. Pinto explains that good scientists have more ideas than they can pursue and are excited to help generate other ideas that others can follow to do good science. Perhaps the more closed responses from others outside their university signify that she could have potential competitors for her particular project. Many times, within a university, it is easier and more fruitful to cooperate than to compete for resources against colleagues. But he points out that it is not uncommon for unrelated researchers to converge on ideas and even results at the same time. For example, in December 9, 2009, issue of *Nature*, four separate papers described various features of the receptor for the plant hormone abscisic acid. Stated another way by Abraham Lincoln, "Books serve to show a man that those original thoughts of his aren't very new at all." So, Pinto advises Ginger, that if she wishes to go into an exciting topic, to expect competition, to work hard, be smart, and do the best research possible. And he adds, "Be speedy, Gonzales." Both Ginger and Pinto are excited and, through Ginger's resourcefulness, they find some new collaborators to venture into this new area of research.

During her discussions, a few faculty members at her university tell her about Dr. Carlos Hutten, who is a peatland ecologist in a small applied department at the same university. They suggest she visit him. Dr. Hutten, an older associate professor, has not published extensively, but is a very pleasant person interested in traditional approaches to ecology. She makes an appointment to visit with him and during their discussion she notices that he behaves decidedly different than everyone else at the university she'd spoken to her about project. After listening to Ginger excitedly tell of her ideas, Dr. Hutten then describes his own interest as being similar and actually encompassing her own. "I've been a peatland ecologist for 40 years. What does Dr. Pinto know about peatlands? I wrote a proposal 25 years ago about this very topic. I knew global climate change was big while everyone else was still doubters. I even suggested that peatlands would be arbiters and bellwethers of global change." Ginger listened carefully while he continued. "I'm very concerned about you entering this field, because you'd be competing with me. In fact, I think Pinto has stolen my research before, which accounts for a good deal of his success. Furthermore, it was because of his theft that my grant proposal wasn't funded." Ginger, now confused, asks what he thinks she should do. She didn't know all these things about Dr. Pinto, and she certainly doesn't want to compete with a professor in her own university; especially since, it seems, these were his ideas all along. Dr. Hutten suggests that she really has just two options. One, he explains, is to pick another topic not associated with peatland ecology. "I'm the peatland ecologist here," he repeats. The other option is that she work directly with him and not with Dr. Pinto. "I could probably get you a teaching assistantship," he suggests.

After her meeting with Dr. Hutten, Ginger is confused and does not know what to do next. She has suddenly lost excitement for her project and has doubts about continuing in Dr. Pinto's lab. How should she proceed now with all this new information?

Discussion questions

1. Should Ginger discuss the issue further with Dr. Pinto and others? Maybe they know more of the history with Drs. Hutten and Pinto. There are at least two sides to every story.

2. Perhaps she should not compete with Dr. Hutten since he was there first? After all, what chance does a PhD student have competing against a professor?

3. Should Ginger consider simply continuing to pursue her ideas for research and surround herself with willing collaborators and competent scientists? After all, it appears that Hutten does not publish very fast or very often. "Don't worry too much about what the competition is doing; worry about controlling your own destiny," Pinto says.

4. It is possible she should switch universities since there may be politics going on where she is currently?

5. Should professors discuss their distaste for one another with students? Should Hutten have criticized Pinto to Ginger? Where does giving information end and gossip start?

Summary

✓ Conflicts of interest occur when one activity or interest could corrupt or disrupt the pursuit of a constraining interest.

✓ Conflicts of commitment occur when opposing commitments in time or energy where one or both will be accomplished suboptimally or not at all.

✓ It is important to disclose potential conflicts as they arise and discuss these with university administrators if there is concern.

✓ Conflicts of interest should also be fully disclosed during funding activities, such as in grant proposal applications.

Chapter 13

What Kind of Research Science World Do We Want?

ABOUT THIS CHAPTER

- The best science can be characterized by "a culture of discipline and an ethic of entrepreneurship."
- Without doing science right, i.e., best ethical practices, it is impossible to do the right science.
- The practice of integrity is irreplaceable in research science.

The ultimate goal of science is to gain and apply knowledge to help people and the world. Research, both basic and applied, seems to be the best way to gain this knowledge and apply it to address practical problems. How most scientists currently go about research is to raise funds to allow them to follow their noses in their areas of interest. Competitive funding allows scientists to then perform the right experiments to address testable hypotheses designed to give a decent certainty that the answers obtained are likely the right or wrong ones. In scientific terms, hypotheses can be rejected or they can fail to be rejected. Through the weight of evidence, we can then discern which answers are likely to be the best ones to answer the questions posed. The knowledge is then extensively vetted to the entire world with special interest from other research scientists in the field for verification, criticism, and if appropriate, tacit acceptance. This knowledge is then applied and refined as it then serves as the basis for more research and development by other scientists and engineers. University scientists, perhaps the largest cohort of scientists worldwide, simultaneously perform research while also training the next generation of scientists.

Research Ethics for Scientists: A Companion for Students, Second Edition. C. Neal Stewart, Jr.
© 2023 John Wiley & Sons Ltd. Published 2023 by John Wiley & Sons Ltd.
Companion website: www.wiley.com/go/stewart/researchethics2

A big part of this training is simply allowing students and postdocs to learn by doing under the supervision or (more accurately) apprenticeship by more established scientists. Essentially, this is the science world we have. In many ways, these components are optimized for efficiency and success. In practice, of course, there are flaws (Ritchie 2020). There are always flaws when people are involved. We live with honest mistakes and judgment errors within the system. Science becomes broken when ethical violations occur, especially in terms of FFP, but bad mentoring and rogue trainees, in my opinion, can be as damaging as some FFP.

At times, we can't disassociate current political and social movements from our laboratory activities. Science has increasingly gotten more political, even though, in the United States at least, both conservative and liberal politicians seem to equally promote public funding of science. But in the past few years, there appears to be distrust within some circles of anything hinting of elitism. And, of course, scientists are highly trained over many years of formal education and then honing their craft within research universities. How could it be any other way?

Of course, the COVID-19 pandemic appeared to exacerbate a trend of growing anti-science sentiments within the general populace. During COVID-19, biomedical science came into the spotlight. I never dreamed that any government scientist would be a household name, but NIH's Anthony Fauci became the face of the pandemic. He was both loved and reviled by various segments of the population, seemingly commensurate with political viewpoints. Vaccines, one of the main tools for societies to survive the pandemic, came under the spotlight at a time where some very vocal people were expressing skepticism of vaccines, including misinformation about the science behind vaccine development. The anti-vax environment greatly enabled by the infamous Andrew Wakefield. He is the discredited scientist whose retracted paper claiming vaccines cause autism and a leader in the anti-vax movement (Deer 2020). But there is more. In the rush to publish coronavirus research, some shortcuts were made in the scientific process that resulted in both FFP and honest error that outpaced the self-correcting system that is science (Pearson 2021). Arguably, the stress of COVID-19 and the "perfect" sociopolitical storm created a science world that few could imagine. Much of it was not pretty. Within these socio-political wars in the pandemic, various scientists weighed in with their expertise. Shockingly, they were attacked in many ways, which included death threats (Nogrady 2021).

Where do we go from here? How can we personally implement improvements in our collective practices as scientists? Is simple refinement of an already efficient process adequate or do we need to demolish the current system and build a fundamentally different science world from scratch?

A culture of discipline and an ethic of entrepreneurship

I would like to suggest that the current system of science has evolved as the most effective one to accomplish the best science. We do not need to create a completely new system to replace it. The system has features that allow science to move relatively quickly while maintaining high quality and rigor. The most tacit feature is that science is populated by smart and disciplined entrepreneurs. At least, these are the people who are currently rewarded and thereby predominantly shape the face of science. They select this system of science and science selects these kinds of people. The meritocracy is then self-sustaining. So, I'd like to end the book on a note of optimism and stress that ideals of science are noble ideals that simply need to be executed with an ethic of integrity. This ethic of integrity combined with noble ideals should result in best practices, many of which were examined in detail in the prior chapters. This ethic will ride out the storms, because it ultimately tells the truth.

In his book *Good to Great*, James Collins (2001) describes common characteristics of companies that were good and went on to become great companies. One of my favorite lines in the book is "When you combine a culture of discipline with an ethic of entrepreneurship, you get the magic alchemy of great performance." I think that this sentence also describes the best scientists who accomplish the best science. You can likely apply this statement to most Nobel Laureates and winners of major science awards – this is how they accomplished great things: self-discipline that yields a disciplined lab setting, which encourages entrepreneurship. Of course, these men and women of science are not perfect and indeed more than one major prize winner has been found in controversies and FFP that could certainly have been avoided. But I think it is safe to say that most of those who have been recognized as the best are the best. And they inspire us to adopt best practices so our own science can improve.

Ethic of entrepreneurship

Entrepreneurs are people who are start and run their own companies. They have boldness and confidence to conceive and act on winning ideas that are then brought into fruition to deliver novel and useful products. Substitute the word "research" for "product" and you have now defined the successful academic scientist. University research, by its nature, is a bottom-up enterprise that rewards the successful entrepreneurs who can sustainably fund and carry out impactful research. Government and industry science, long known for its top-down management, is becoming more entrepreneurial in nature in which the best ideas and teams are sought in competitive endeavors. There is a convergence on the entrepreneurial model because it is indeed effective in providing sound research to address rapidly changing problems. As we recall the philosophical underpinnings of research ethics in Chapter 1, although it was not called the "ethic of entrepreneurship" that is exactly what egoism represents: people doing what is in their own best interest. Entrepreneurs must focus on the narrow set of activities ensuring that their companies or activities will succeed, which, by definition, is in their own self-interest. One of the surest routes to success is discipline: self-discipline.

Culture of discipline

The best scientists I know are extremely disciplined people. Forget about the stories James Watson tells in his book *The Double Helix* (1968), where he states that he and Francis Crick arrived in the laboratory at 10:00 a.m. and then knocked off at 2:00 p.m. to have fun. They were intently focused on elucidating the structure of DNA and on that scientific question alone. Having some down time to ponder was part of the plan at the time, but in the rest of their careers they were extraordinarily focused and disciplined in all the ways you might imagine scientists being disciplined. That is the only way to have sustained productivity. The best scientists are disciplined to think, read, design experiments, and write grant proposals and papers. The best scientists I know are disciplined to be competent and sustain competence over the long haul. They are disciplined to treat their trainees well and to give them what they need to, in turn, succeed in science. The best scientists plan their time schedules. They make goals for their days, weeks, and years. The have disciplined thoughts and actions. The opposite of discipline is that which leads to ethical problems. Even in the cases of Type A personalities such as we saw in Hwang and Schön, they did not have the patience and discipline to let the scientific process take its course. They made undisciplined shortcuts by fabricating data. It might very well be that the root of most research misconduct is a lack of discipline of one sort of the other. Therefore, the science world I want is one that is described as "a culture of discipline and the ethic of entrepreneurship."

Judge yourself

✓ Are you a disciplined person? Can you channel discipline to produce fruitful science?
✓ Entrepreneurs don't tackle other people's projects but create their own. How entrepreneurial are you?
✓ Do you want to be the best or is being good simply good enough?

Too much pressure?

There exists an argument that the pressure of funding and the culture of science that focuses on competition, an academic "up or out" tenure system, and not enough research funds is the cause of problems in research integrity (Munck 1997; Ritchie 2020). I think research integrity is largely a human problem that can be solved, at least partially, by education and the courage to hold people accountable. I'm thoroughly convinced that eliminating competition and giving all scientists virtually unlimited funding would not to improve the quality of science. I think substantially more available money for research would simply increase the amount of mediocre research and not make the best research better. Indeed, in such a situation there would be more mediocre scientists training students to be mediocre, which would increase the population of mediocre scientists producing more mediocre science. I once posed the following question to a group of administrators. If you gave your least productive researchers all the money they wished to have, would it elevate their productivity to that of your most productive faculty? No one answered in the affirmative. More funding, while it might be somewhat helpful, will not substitute for competence and discipline in the researcher. But what about the temptation to cheat to win grants? Yes, of course, there are always temptations and the drive for survival. In many ways, there are some practical aspects of the academic system, at least in the United States, that might lead to problems. But I believe the problems are manageable.

I think there are immense pressures for scientists to become one-dimensional workaholics in order to stay competitive in research and to win tenure. To some degree, this pressure is temporary. It begins in graduate school and tends to peak during the tenure-getting years leading up to the "big decision" whether a university will retain an assistant professor by granting tenure and promotion to associate professor or firing the person. With so much invested in training to get the PhD, doing postdoc stints, and then toiling for years to get tenure, there is a lot at stake for the young scientist who might not earn tenure. I am no fan of the tenure system. I've now gotten tenure three times and walked away from tenure twice. I am, however, currently tenured. Certainly, it adds undue pressure in the early years, which might contribute to compromised research integrity, but then tenure protects the deadwood, i.e., unproductive scientists, who do not deserve to be protected. I would rather see a system of five-year contracts in which outstanding scientists would enjoy perpetual five-year contract extensions and low performers would experience contract terminations. This would introduce higher accountability while taking a bit of the pressure off in the assistant professor years.

In a dramatic case of the tenure-decision pressure cooker, Amy Bishop murdered her department head and two of her colleagues at the University of Alabama-Huntsville in February 2010. Certainly the personal stress associated with the tenure system played a role, but further analysis of her background shows questionable behaviors and a history of violence even prior to her negative tenure decision. Unfortunately, being denied tenure is the "scarlet letter" of science, branding one as untouchable.

The consensus among US academia is that the tenure system will never be eliminated. If an institution eliminated tenure, it would lose its competitive edge to recruit the best faculty. The non-tenure-system islands in a sea of tenure would sink. Detailed background checks conducted prior to hiring scientists may be used to avoid another Bishop-like disaster. To deal with deadwood institutions may – as my own university has done recently – implement post-tenure review systems. I recently was subjected to a relatively painless post-tenure review and passed. I am unaware of how effective this system is to "encourage" nonproductive faculty members to retire as the system is quite new. As expected, post-tenure review is not so popular with people accustomed to assurance of a "job-for-life."

The other unchangeable system, in my opinion, is that of competitive grants. Researchers, such as those in biomedical areas, lament decreased funding rates in recent years, but this is because there are probably more biomedical scientists being trained than the funding system can bear. Funding is a product of real needs in various parts of society, politics, and economics. We see politics and societal pressures to land a person on the moon driving the huge budgets on NASA in the 1960s that carried over to the 1970s (Figure 9.1). Here also we see that the oil crisis drove increased research expenditures in bioenergy in the 1980s. This dynamic change of priorities is part of life that scientists should expect and embrace. The most successful scientists love to do research and will adjust their interests to some-what track future funding cycles and trends. Part of staying competent is embracing change.

As Billy Joel (yes, the musician) observed, "I am, as I've said, merely competent. But in the age of incompetence, that makes me extraordinary." I'm not convinced the same can be said of many scientists. In fact increasing accountability should also increase competence. Increasing grant competitiveness in one discipline coupled with more opportunities in other disciplines should simply shift the areas of science in which people are being trained. History shows us this is how it works. It does no good to bemoan that my favorite area X is no longer being funded; therefore, there is something wrong with the system. When we realize the funding system is a social/political/economic chimera, we should no longer feel the need to make up excuses why the system is flawed. I recall once attending a "research presentation" of a scientist who was on a speaking circuit of sorts, not for his research, but to reassure an audience of like-minded people that funding agencies have lost their missions, etc., which is why nobody in the room was winning as many grants as they did previously. Of course, the fact of the matter was that this particular speaker and much of the audience were simply no longer competitive in the grants world. The world of science had past them by as funding priorities had changed and those scientists, including the speaker, had failed to adapt. Certainly, there are myriad avenues to funding beyond competitive grants, such as through companies and foundations. I've enjoyed these too, but I've observed in my own career that competitive grants are one of the primary means by which I remain more competitive and competent in science. Writing full-blown proposals keeps me sharp. Easy money typically doesn't demand the stringency or rigor in planning research that does competitive research. For that reason I think that a skewed demand and supply spectrum is probably good for science. So, let's take a look at some key features that we can address that will give us the biggest payoff with regards to quality of science research and integrity.

Integrity awareness through ethics education

By far, the best thing that established scientists can do for young scientists is to teach ethics in research. Many mentors do just that, but now is the time to establish formalized mandatory ethics training in graduate education at the institution level (Titus and Bosch 2010). I love the one-credit hour research ethics I co-teach with other faculty members. Student evaluations concur that the course makes a big difference in their awareness of ethical issues in being a scientist. In fact, many students wish the course was longer and had more credit hours. A case-study and discussion-based course with 15 students or fewer is rewarding for all who are involved (Stewart and Edwards 2008). In addition, there is also a need for research ethics orientation for postdocs and new faculty as well. In this sort of educational program, we should focus on the ethical development of the scientist and best practices that give best results. I would love to someday develop an undergraduate course on research integrity, since researchers are starting younger and they ought to better understand the system of science prior to beginning their graduate training.

Accountability

The increased accountability that is most notable is fiscal accountability from granting agencies and also, perhaps greater reporting requirements. The accountability that I propose is probably better described as joint responsibility for ensuring research integrity. Thus, the focus should be on the science itself and not the scientist. There is a school of thought, usually not overt, that borders on clubism. What do I mean by that? Simply, that scientists will protect other scientists, especially those in their "club," be it gender, political persuasion, department, commodity, interest, scientific society, or discipline. I think this protectionism and loyalty to friends, preferentially, than to science or truth is prevalent, and it is fatal in the cause of maximizing research integrity. Of course, loyalties to people run deep, and there is enduring hope of reciprocal loyalty as well. Therefore, we don't want to "call out" colleagues when they err. It is important to keep this in mind when handling manuscripts and grant proposals. I know that in publishing we may be slow to point out possible ethical missteps, and then editors are sometimes reticent to act on them. I worry about all the new journals being created with the commensurate "need" to publish papers. In the urgency to publish, it is not beneficial if authors, peer reviewers, and editors cut corners, especially editors. Many journals and scientific societies do not have any ethics statements or procedures on how they will handle breaches of ethics. Clearly, editors, especially, must be proactive to maintain "purity" and integrity in science. There is a prevailing view among some scientists that journals exist so that scientists have venues to publish to advance their careers and their students' interests. That is a perverted perspective. Journals exist to publish sound science. There is no shortage of self-help/virtual mentor books that are centered on giving advice about how to advance one's career in science. While there is nothing wrong with career advancement per se, that should not be the principal motivation for pursuing a career in science. The main goal should be to advance knowledge. In my opinion, it is in the outlook and motivation where people sometimes go wrong. When we are so set on advancing our own careers and the careers of our pals, we can easily lose sight of research integrity.

Truth will win

In the aftermath of the COVID-19 pandemic and blatant attacks on science from politicians and their followers, being a scientist can be discouraging. But, in this regard, today in the United States it is not as bad the USSR was in the 1940s through the early 1960s when Trofim Lysenko was a thought leader in plant breeding and agronomy. His warped views on genetics became politically attractive to Josef Stalin and the Communist Party as a means to improve agricultural output. Lysenko was appointed at the Director of the Institute of Genetics in the USSR Academy of Science, which was ironic given that he didn't believe that genes played any role in inheritance. Under his power and with government blessing, he discredited and persecuted those scientists who didn't share his views, and essentially put Russian genetics in a hole that took decades to climb from. His policies also led to famines and poverty. Many scientists paid a dear price for crossing Lysenko, such as his mentor Nikolai Vavilov who starved to death in prison in 1943 (Offord 2021). Eventually, truth won out, but at great costs. This is a hard lesson for science and for leaders of countries who decide to make scientifically illiterate decisions in spite of sound scientific advice. For our part, we as scientists are ultimately responsible for telling the truth and maintaining the system of scientific research that is a proven winner.

We scientists

Incremental science advancements should result in sound publications and be rewarded with grants. Incremental advancement in a specific discipline toward solution of a specific problem is a worthy goal. Science is a meritocracy. Breakthroughs are wonderful, but they should not be the objective each morning when the lab door is unlocked. Breakthroughs are often serendipitous, occurring when we least expect them. We need to temper our expectations and feel good about doing sound research that leads to solid advancement. Likewise, prizes are also wonderful, but seeking to win a Nobel should never be the end-goal of scientists. Seeking truth is the ultimate goal. The pursuit of sound incremental science is the way to change the world for the better.

We also need to take the time to train the next generation of scientists in best practices; not just how to get ahead or how to win the grant, but how to truly succeed in science. This is the focus of egoism – a long view for doing what is in one's own self-interest need not be self-centered. I like to think that by advancing science and by being flexible, that there will be no shortage of problems for which to apply scientific research. Mike Tyson (yes, the boxer) said, "Everyone has a plan until they get punched in the mouth." All PhD students plan to continue their favorite topic of research until they get punched in the mouth (that it is not very fundable or applicable as they thought).

We also need to expect great things from the next generation of scientists. Is that goal at odds with respecting incremental accomplishments? In no way. I have relatively few regrets in my career that has spanned over 30 years. I regret not being rigorous enough with students, postdocs, and myself. I regret those times when we've taken shortcuts and not engaged in best practices. I regret those times when I've been lazy in not running strong enough when the finish line is in sight. I regret taking such a long time to really pay attention to research ethics. From teaching a simple discussion course on ethics in our departments, to being both kind and demanding with our students, to being a rigorous reviewer of journal articles: this is the kind of science world that I want. This is the kind of science world that society will respect and reward.

Summary

✓ Science as a way of life is based on discipline and excellence to pursue rigor.
✓ Entitlements don't encourage discipline, rigor, or excellence.
✓ Education on ethics and RCR training is key to becoming better scientists with a culture of integrity.

References

500 Women Scientists. 2021. Eric Lander is not the ideal choice for Presidential Science Advisor. *Scientific American*, January 21, 2012. https://www.scientificamerican.com/article/eric-lander-is-not-the-ideal-choice-for-presidential-science-adviser/.

Abbott, A. 2022. Max Planck's cherished autonomy questioned. *Nature* **606**:632–633.

Adams, D. and Schwendinger, K. 2020. Opinion: disclosures scientists must make of foreign ties. *The Scientist*, February 12, 2020.

Akst, J. 2010. I hate your paper. *The Scientist* **24**:36–41.

Allen, W. 1985. *Habitat Suitability Index Models: Swamp Rabbit.* US Fish and Wildlife Service, National Wetlands Research Center, Lafayette, LA. Biological Report 82 (10.107). 20pp.

Allen, J. 2008. Can of worms. *On Wisconsin Magazine*, Spring.

Anonymous. 2005. Timeline of a controversy. Retrieved 2009 from *Nature News*: http://www.nature.com/news/2005/051219/full/news051219-3.html.

Anonymous 2006. Beautification and fraud. *Nature Cell Biology* **8**:101–102.

Anonymous 2009. Editorial: credit where credit is overdue. *Nature Biotechnology* **27**:579.

Anonymous 2010. The sequence explosion. *Nature* **464**:670–671.

Associated Press. 2009. Senator's affair highlights capitol temptations. *Knoxville News Sentinel*, August 10, 2009, p. A4.

Augenbraun, E. 2008. Letter to the editor. *The Scientist* **22**(5):17.

Barnbaum, D. R. and Byron, M. 2001. *Research Ethics: Text and Readings.* Prentice Hall, Upper Saddle River, NJ.

Bauerlein, M., Gad-el-Hak, M., Grody, W., McKelvey, B. and Trimble, S. W. 2010. We must stop the avalanche of low-quality research. *The Chronicle of Higher Education.* Published Online: http://chronicle.com/article/We-Must-Stop-the-Avalanche-of/65890/.

Berlin, L. 2009. Plagiarism, salami slicing, and Labachevsky. *Skeletal Radiology* **38**:1–4.

Research Ethics for Scientists: A Companion for Students, Second Edition. C. Neal Stewart, Jr.
© 2023 John Wiley & Sons Ltd. Published 2023 by John Wiley & Sons Ltd.
Companion website: www.wiley.com/go/stewart/researchethics2

Bordons, M. and Gomez, I. 2000. Collaboration networks in science. Pp. 197–213. Published in Cronin, B. and Atkins, H. B. (eds), *The Web of Knowledge*. American Society for Information Science, Medford, NJ.

Bouter, L. M., Tijdink, J., Axelssen, N., Martinson, B. C. and ter Riet, G. 2016. Raning major and minor research misbehaviors: results from a survey among the participants of four World Conferences on Research Integrity. *Research Integrity and Peer Review* **1**:7.

Brainard, J. 2022. Journal declares and end to accepting or rejecting papers. *Science* **378**:346.

Breitling, L. P. 2005. Misconduct: pressure to achieve corrodes ideals. *Nature* **436**:626.

Brown, B. and Nguyen, A. L. 2021. In case of death. *Science* **372**:1358.

Brumfiel, G. 2008. Physicists all aflutter about data photographed at conference. *Nature* **455**:7.

Brumfiel, G. 2009. Breaking the convention? *Nature* **459**: 1050–1051.

Bush, V. 1945. *Science – the Endless Frontier*. United States Government Printing Office, Washington, DC.

Butler, D. 2010. Journals step up plagiarism policing. *Nature* **466**:167.

Cardenuto, J. P. and Rocha, A. 2022. Benchmarking scientific image forgery detectors. *Scientific and Engineering Ethics* **28**:35.

Carlson, S. 2010. Lab notebook tips from a patent litigator. *Genetic Engineering News* **30**(1):10–11.

Chapman, J. A. and Feldhamer, G. 1981. Sylvilagus aquaticus. *Mammalian Species* 151:1–4.

Chapman, J. A., Hockman, J. G. and Edwards, W. R. 1982. Cottontails. Pp. 83–123. Published in Chapman, J. A. and Feldhamer, G. A. (eds), *Wild Mammals of North America: Biology, Management, Economics*. Johns Hopkins University Press, Baltimore, MD. 1147pp.

Cohen, J. 2011. Dispute of lab notebooks lands researcher in jail. *Science* **334**:189–1190.

Collberg, C., Kobourov, S., Louie, J. and Slattery, T. 2003. Retrieved 2010 from *Splat: A System for Self-Plagiarism Detection*, Department of Computer Science, University of Arizona: http://splat.cs.arizona.edu/icwi_plag.pdf.

Collins, J. 2001. *Good to Great*. Harper Business, New York.

Comstock, G. L. (Ed). 2002. *Life Science Ethics*. Iowa State Press, Ames, IA.

Couzin, J. 2006a. U.S. rules on accounting for grants amounts to more than a hill of beans. *Science* **311**:168–169.

References

Couzin, J. 2006b. Stem cells … and how the problems eluded peer reviewers and editors. *Science* **311**:23–24.

Couzin, J. 2006c. Truth and consequences. *Science* **313**:1222–1226.

Couzin-Frankel, J. 2010. The legacy plan. *Science* **329**:135–137.

Croll, R. P. 1984. The non-contributing author: an issue of credit and responsibility. *Perspectives in Biology and Medicine* **27**:401–407.

Curry, A. 2021. Max Planck director loses post after probe of misconduct. *Science* **374**:671.

Cyranosaki, D. 2006. Rise and fall. Retrieved 2010 from *Nature News*: http://www.nature.com/news/2006/060111/full/news060109-8.html.

Deer, B. 2020. *The Doctor who Fooled the World: Andrew Wakefield's War on Vaccines*. Scribe.

Else, H. and Van Noorden, R. 2021. The battle against paper mills. *Nature* **591**:516–519.

Errami, M. and Garner, H. 2008. A tale of two citations. *Nature* **451**:397–399.

Fischbach, R. L. and Plaza, J. 2003. The case of the promising new drug. Retrieved 2009 from *Responsible Conduct of Research*, Columbia University: http://ccnmtl.columbia.edu/projects/rcr/rcr_conflicts/case/index.html.

Fowler, A. and Kissell, R. E. Jr. 2007. Winter relative abundance and habitat associations of swamp rabbits in eastern Arkansas. *Southeastern Naturalist* **6**:247–258.

Gert, B. 1997. Morality and scientific research. Pp. 20–33. Published in Elliott, D. and Stern, J. E. (eds), *Research Ethics*. University Press of New England, Hanover, NH.

Gewin, V. 2021. How to blow the whistle on an academic bully. *Nature* **593**:299–301.

Green, L. 2005. Reviewing the scourge of self-plagiarism. *M/C Journal* **8**(5). Retrieved 2010 from http://journal.media-culture.org.au/0510/07-green.php.

Grudniewicz, A. et al. 2019. Predatory journals: no definition, no defence. *Nature* **576**:210–212.

Gu, J., Wang, X., Li, C., Zhao, J., Fu, W., Liang, G. and Qiu, J. 2022. AI-enabled image fraud in scientific publications. *Patterns* **3**(7):100511.

Gunsalus, C. K. 1998. How to blow the whistle and still have a career afterwards. *Science and Engineering Ethics* **4**:51–64.

Handwerker, H. 2010. Impact factor blues. *European Journal of Pain* **14**:3–4.

Heim, R. and Tsien, R. Y. 1996. Engineering green fluorescent protein for improved brightness, longer wavelengths and fluorescence resonance energy transfer. *Current Biology* **6**:178–182.

Herman, I. P. 2007. Following the law. *Nature* **445**:228.

Hirsch, J. E. 2005. An index to quantify an individual's research output. *Proceedings of the National Academy of Sciences USA* **102**: 16569–16572.

Judson, H. F. 2004. *The Great Betrayal: Fraud in Science*. Harcourt Inc., Orlando, FL.

Kjolhaug, M. S. and Woolf, A. 1988. Home range of the swamp rabbit in southern Illinois. *Journal of Mammalogy* 69:194–197.

Koocher, G. and Keith-Spiegel, P. 2010. Peers nip misconduct in the bud. *Nature* **466**:438–440.

Louis, K. S., Blumenthal, D., Gluck, M. E. and Soto, M. A. 1989. Entrepreneurs in academe: an exploration of behaviors among life scientists. *Administrative Science Quarterly* **34**:110–131.

Lynch, C. 2008. Big data: how do your data grow? *Nature* **455**:28–29.

Macrina, F. L. 2005. *Scientific Integrity*. Third edn. American Society for Microbiology Press, Washington, DC.

Malmgren, R., Ottino, J. and Nunes Amaral, L. 2010. The role of mentorship in protégé performance. *Nature* **465**:622–626.

Martin, B. 1992. Scientific fraud and the power structure of science. *Prometheus* **10**:83–98.

Martinson, B. C., Anderson, M. S. and De Vries, R. 2005. Scientists behaving badly. *Nature* **435**:737–738.

May, M. 2009. Sharing the wealth of data. *Scientific American Worldview*, pp. 88–93.

McCook, A. 2009. Life after fraud. *The Scientist* **23**(7):28–33.

McGee, G. 2007. Me first. *The Scientist* **12**(9):28.

Mervis, J. 2022. Fraud charges crumble in China Initiative cases. *Science* **377**:1478.

Munck, A. 1997. Examples of scientific misconduct. Pp. 31–33. Published in Elliott, D. and Stern, J. E. (eds), *Research Ethics*. University Press of New England, Hanover, NH.

National Academy of Sciences 2009. *Ensuring the Integrity, Accessibility, and Stewardship of Research Data in the Digital Age*. National Academies Press, Washington, DC.

Nelson, B. 2009. Empty archives. *Nature* **461**:160–163.

Nogrady, B. 2021. Scientists under attack. *Nature* **598**:250–253.

Normile, D. 2009. *Science* retracts discredited paper; bitter patent dispute continues. *Science* **324**:450–451.

O'Grady, C. 2021. Image sleuth faces legal threats. *Science* **372**: 1021–1022.

Offord, C. 2021. Stamping out science, 1948. *The Scientist*, May 1, 2021.

Pearson, H. 2009. Being Bob Langer. *Nature* **458**:22–24.

Pearson, H. 2021. How COVID broke the evidence pipeline. *Nature* **593**:182–185.

Pool, R. 1997. More squabbling over unbelievable result. Pp. 117–138. Published in Elliott, D. and Stern, J. E. (eds), *Research Ethics*. University Press of New England, Hanover, NH.

Prasher, D. C., Eckenrode, V. K., Ward, W. W., Prendergrast, F. G. and Cormier, M. J. 1992. Primary structure of the *Aequorea victoria* green fluorescent protein. *Gene* **111**:229–233.

Reis, R. 1999. Choosing a research topic. *The Chronicle of Higher Education*. Published Online: http://chronicle.com/article/Choosing-a-Research-Topic/45641/.

Ritchie, S. 2020. *Science Fictions: How Fraud, Bias, Negligence, and Hype Undermine the Search for Truth*. Metropolitan Books, New York.

Rosser, M. and Yamada, K. M. 2004. What's in a picture? The temptation of image manipulation. *Journal of Cell Biology* **166**:11–15.

Shamoo, A. E. and Resnik, D. B. 2003. *Responsible Conduct of Research*. Oxford University Press, New York.

Shen, H. 2020. Seeing double. *Nature* **581**:132–136.

Spier, R. 2002. The history of the peer-review process. *Trends in Biotechnology* **20**:357–358.

Stapel, D. 2016. Derailment: faking science: a true story of academic fraud (translated into English by N.J.L. Brown). Published and copyrighted by the authors.

Stewart, C. N. Jr. and Edwards, J. E. 2008. How to teach research ethics. *The Scientist* **22**(2):27.

Thorisson, G. A. 2009. Accreditation and attribution in data sharing. *Nature Biotechnology* **27**:984–985.

Titus, S. and Bosch, X. 2010. Tie funding to research integrity. *Nature* **466**:436–438.

Toffelson, J. and Van Noorden, R. 2022. US government reveals bug changes to open-access policy. *Nature* **609**:234–235.

Van Noorden, R. 2010. A profusion of measures. *Nature* **465**:864–866.

Van Noorden, R. 2022. Journals adopt AI to spot duplicated images in manuscripts. *Nature* **601**:14–15.

Vogel, G. 2011. Jan Hendrick Schön loses his PhD. ScienceInsider https://www.science.org/content/article/jan-hendrik-sch-n-loses-his-phd.

Waldrop, M. 2008. Wikiomics. *Nature* **455**:22–25.

Wang, K. 2005. Education and penalties are key to tackling misconduct. *Nature* **436**:626.

Warren, W. L. and Cao, S. A. 2009. Certainty not required for inventorship. *Genetic Engineering News* **29**(10):12.

Watland, A. M., Schauber, E. M. and Woolf, A. 2007. Translocation of swamp rabbits in eastern Illinois. *Southeastern Naturalist* **6**:259–270.

Watson, J. D. 1968. *The Double Helix: a Personal Account of the Discovery of the Structure of DNA*. New American Library, New York.

Weale, S. 2018. Top cancer genetics professor quits job over bullying allegations. *The Guardian*, July 17, 2018. https://www.theguardian.com/society/2018/jul/17/top-cancer-genetics-professor-quits-job-over-bullying-allegations.

Weaver, W. 1948. Science and complexity. *American Scientist* **36**:536–544.

Weil, V. and Arzbaecher, R. 1997. Relationships in laboratories and research communities. Pp. 69–90. Published in Elliott, D. and Stern, J. E. (eds), *Research Ethics*. University Press of New England, Hanover, NH.

Weston, A. 2002. *A Practical Companion to Ethics*. Second edn. Oxford University Press, New York.

Wilson, J. R. 2002. Responsible authorship and peer review. *Science and Engineering Ethics* **8**:155–174.

Winter, S. 2010. Former Wisconsin researcher sentenced for misconduct. *BioTechniques Newsletter*, September 23, 2010. http://www.biotechniques.com/news/Former-Wisconsin-researcher-sentenced-for-misconduct/biotechniques-302891.html?utm_source=BioTechniques+Newsletters+%2526+e-Alerts&utm_campaign=d9d482b24e-BioTechniques_Weekly&utm_medium=email (last accessed November 17, 2010).

Woolf, P. 1997. Pressure to publish and fraud in science. Pp. 141–145. Published in Elliott, D. and Stern, J. E. (eds), *Research Ethics*. University Press of New England, Hanover, NH.

Woolston, C. 2020. Postdocs under pressure: 'Can I even do this any more?'. *Nature* **587**:689–692.

Wuchty, S., Jones, B. F. and Uzzi, B. 2007. The increasing dominance of teams in production of knowledge. *Science* **316**:1036–1039.

Zhang, Y. 2010. Chinese journal finds 31% of submissions plagiarized. *Nature* **467**:153.

Index

Research Ethics for Scientists: A Companion for Students, Second Edition. C. Neal Stewart, Jr.
© 2023 John Wiley & Sons Ltd. Published 2023 by John Wiley & Sons Ltd.
Companion website: www.wiley.com/go/stewart/researchethics2

Index

Printed and bound by CPI Group (UK) Ltd, Croydon, CR0 4YY

06/07/2023

03233224-0001